Applied Physical Geography

GEOSYSTEMS IN THE LABORATORY

Tenth Edition

Charles E. Thomsen

American River College

Robert W. Christopherson

American River College - Emeritus

Applied Physical Geography

GEOSYSTEMS IN THE LABORATORY

Tenth Edition

Charles E. Thomsen

American River College

Robert W. Christopherson

American River College - Emeritus

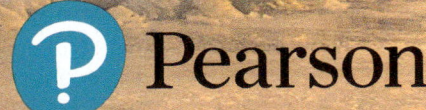

330 Hudson Street, NY NY 10013

Executive Editor, Geosciences Courseware: Christian Botting
Courseware Director, Portfolio Management: Beth Wilbur
Managing Producer, Science: Mike Early
Content Producer: Anton Yakovlev
Production Management: Jeanine Furino, Cenveo Publisher Services
Editorial Assistant, Geosciences Courseware: Emily Bornhop
Executive Field Marketing Manager: Mary Salzman
Illustrators: Lachina and International Mapping Associates
Senior Content Producer: Tim Hainley
Cover Photo: *Island in Lake Baikal*, by Chalermkiat Deedokmai/Getty Images

Photo Credits: Opener Images: Chapters 1-11, 30: FotoVoyager/E+/Getty Images. Chapters 12-18: Michael & Jennifer Lewis/Getty Images. Chapters 19-27: David Noton Photography/Alamy. Chapters 28-29: Paul Baggaley/Getty Images.

Library of Congress Cataloging-in-Publication Data

Names: Thomsen, Charles E., author. | Christopherson, Robert W., author.
Title: Applied physical geography : geosystems in the laboratory / Charles E.
   Thomsen, American River College, Robert W. Christopherson, American River
   College--Emeritus.
Other titles: Geosystems
Description: Tenth edition. | Boston : Pearson, [2017] | Title of previous
   edition: Geosystems : an introduction to physical geography. | Previous
   edition entered under: Christopherson, Robert W.
Identifiers: LCCN 2017024219 | ISBN 9780134686363 (alk. paper) | ISBN
   0134686365 (alk. paper)
Subjects: LCSH: Physical geography.
Classification: LCC GB54.5 .C48 2017 | DDC 910/.02--dc23
LC record available at https://lccn.loc.gov/2017024219

ISBN-13: 978-0-134-68636-3
ISBN-10: 0-134-68636-5

CONTENTS

CORRELATION GRID

	Geosystems 10e	Elemental Geosystems 9e	Geosystems Core 1e
1. Latitude and Longitude	1	1	I
2. The Geographic Grid and Time	1	1	I
3. Directions and Compass Readings	1	1	I
4. Map Projections, Map Reading, and Interpretation	1	1	I
5. Earth—Sun Relationships and Daylength	2	2	1
6. Insolation and Seasons	2	2	1
7. Temperature Concepts	4	3	2
8. Temperature Patterns	4	3	2
9. Temperature Maps	4	3	2
10. Earth's Atmosphere: Temperature and Pressure Profiles	3	2	1
11. Earth's Atmosphere: Pressure and Wind Patterns	5	4	3
12. Atmospheric Humidity	6	5	4
13. Stability and Atmospheric Processes	6	5	4
14. Weather Maps	7	5	4
15. Midlatitude Cyclones and Hurricanes	7	5	4
16. Water Balance and Water Resources	8	6	5
17. Global Climate Systems	9	7	6
18. Climate Change	10	8	7
19. Plate Tectonics: Global Patterns	11, 12	9	8
20. Plate Tectonics: Faulting and Volcanism	12	10	8
21. The Rock Cycle and Rock Identification	11	9	8
22. Recurrence Intervals for Natural Events	14	12	10
23. Contours and Topographic Maps	1	1	I
24. Topographic Analysis: Fluvial Geomorphology	14	12	10
25. Topographic Analysis: Glacial Geomorphology	17	14	12
26. Topographic Analysis: Coastal and Arid Geomorphology and Sea-Level Rise	15, 16	13	11
27. Topographic Analysis: Karst Landscapes	13	11	9
28. Soils	18	15	13
29. Biomes: Analyzing Global Terrestrial Ecosystems	19, 20	16, 17	14
30. An Introduction to Geographic Information Systems	1	1	I

Welcome to *Applied Physical Geography: Geosystems in the Laboratory*! During the tenth edition of this lab manual, important scientific research and global trends will unfold. Physical geography, as the essential Earth systems spatial science, integrates the latest information and discoveries from around the globe. This is an exciting time to be taking a lab class in which many of the principles of physical geography are applied in specific exercises. These exercises give you an opportunity to work with practices and tools of geographic analysis.

For example, imagine that toward the end of this century, the climate of Illinois in the summer will be more like that of present-day east Texas; that's the climate change forecast! You get to work with this forecast and its implications in Lab Exercise 18.

Our goal in this laboratory manual is to help you use geographic principles, methods, and tools to better relate your physical geography text to understanding Earth's systems. The sources for figures and materials used in this manual are *Geosystems: An Introduction to Physical Geography*, by Robert W. Christopherson and Ginger H. Birkland, *Elemental Geosystems*, also by Christopherson and Birkland, and *Geosystems Core*, by Robert W. Christopherson, Stephen F. Cunha, Charles E. Thomsen, and Ginger H. Birkland, all published by Pearson. This lab manual does not follow the specific chapter sequence of these texts.

New to the 10th Edition

- All figures, including topographic maps, have been extensively revised and are now presented in full-color. The 10th edition includes over 100 updated color figures, 20 maps, and eight new aerial images.

- The full-page topographic maps include Quick Response (QR) code links to Google Earth virtual tours of the location. Most of the topographic maps are also paired with high-resolution aerial imagery. Students can now view and explore many locations via topographic maps, aerial imagery, and Google Earth.

- Exercises have been updated with the latest 2016 and 2017 data.

- Every lab includes a mobile-friendly QR link to a PreLab video that reviews the major goals, concepts, and themes of the lab exercise. These PreLab videos save teachers prep time and help students prioritize the most important concepts and skills.

- Mobile-friendly QR codes link select lab exercises to 37 online animations and videos, two *Mobile Field Trip* videos, and three *Project Condor* Quadcopter videos, presenting students with just-in-time review opportunities and real world case studies as they work through lab exercises.

The Science of Geography

Geography (from *geo*, "earth," and *graphein*, "to write") is the science that studies the interdependence among geographic areas, natural systems, society, and cultural activities over space. As a discipline that synthesizes knowledge from many fields, geography integrates spatial elements to form a coherent picture of Earth. The term *spatial* refers to the nature and character of physical space: to measurements, relations, locations, and the distributions of things. Geography is governed by a method—a spatial approach—rather than by a specific body of knowledge.

Physical geography, therefore, centers on the spatial analysis of all the physical elements and processes that comprise the environment: energy, air, water, weather, climate, landforms, soils, animals, plants, and Earth itself. Physical geography, as a spatial human-Earth science, is in a unique position among sciences to synthesize and integrate the great physical and cultural diversity facing us. Aspects of physical geography requiring spatial analysis are in the news daily, for these are dramatic times relative to human-Earth relations and spatial change.

As a science, geography uses the scientific method. Chapter 1 of the *Geosystems* texts briefly discusses this approach to problem solving. As you work through the laboratory manual, keep in mind this method of analyzing a problem, gathering data, organizing thought, and achieving the discovery of important principles.

Each lab exercise will take varying lengths of time, depending on the emphasis taken in lab. Each lab exercise is organized in the following sequence:

- **Introduction**
  A brief overview of the subject, key terms, concepts involved in the exercise.

- **PreLab Videos**
  Quick Response (QR) codes integrated into introductions lead students to online PreLab Videos, which review the major concepts and goals of each lab exercise.

- **Key Terms and Concepts**
  Principal terms and concepts used in the exercise that are defined in the glossary and that you may want to check in a physical geography textbook.

- **Key Learning Concepts**
  The skills and knowledge you should master upon completion of the exercise.

- **Materials/Sources Needed**
  Items that you will need to complete the exercise.

- **Lab Exercise and Activities**
  These headings denote parts of the exercise and include specific numbered questions and activities to complete.

This laboratory manual is designed for maximum utility—an asset to your learning activities. The standard symbols used on topographic maps appear inside the front cover. Inside the back cover is a map of a modified Köppen climate classification system, which identifies causal elements that produce the pattern of world climates. The back cover features a map of global biomes as Earth's terrestrial ecosystems. A fold-out flap, which you can deploy while you are working with the manual, features common metric-English conversions to assist you in converting between the two systems. The outside portion of the fold-out flap presents the four classes of map projections.

Note: The exercises in this manual are set up in a manner sensitive to the fact that many U.S. students still use English units, whereas Canadian students use metric measures. Conversions are given throughout the manual wherever appropriate. Your instructor will guide you as to which system of units to use in the exercises. The graphs and other items are designed to use either system although metric is preferred.

The initial laboratory exercise, called the *Prologue Lab*, is unique to this lab manual. The Prologue Lab contains some activities that last throughout the term and will not be completed until the last class, acting as threads through the lab and helping you appreciate the relevancy of geography. We have also included a section with Google Earth™ activities. If you have not used this wonderful tool, we hope you find it useful and fascinating. If you have used Google Earth before, we hope you enjoy this new use for it. Some of these tasks work best as group activities, as guided by your instructor.

In order to use the Quick Response (QR) codes, you will need a QR reader app on your mobile device. Apple's App Store and Google Play both have several free options. The Google Earth KMZ files require Google Earth to be installed, either the desktop or mobile version. Both desktop and mobile versions can be downloaded from http://www.google.com/earth/. The Google Earth website also has information about installing the programs, as well as tutorials on the programs.

Preface to the Instructor

Applied Physical Geography: Geosystems in the Laboratory is for the student taking an introductory physical geography course, either concurrently or previously. Included are background materials, brief text explanations, and figures from the *Geosystems, Elemental Geosystems,* and *Geosystems Core* texts, although this laboratory manual can be used with other current textbooks on physical geography.

A separate and complete *Answer Key* for this laboratory manual is available to adopters of this lab manual. Ask your local Pearson representative or download it from www.pearsonhighered.com/irc.

Contact your Pearson local rep for examination copies of *Geosystems, Elemental Geosystems, Geosystems Core*, and *Applied Physical Geography: Geosystems in the Laboratory.*

Course Equipment: We made a deliberate effort to write a manual requiring a minimum of equipment to keep your costs low. Students should provide their own atlas, pencils, colored pencils and pens, metric-English ruler, protractor, calculator, and dictionary. Along with the supplies and equipment you have on hand for labs to supplement this manual, the following items are needed: a sling psychrometer, a compass, weather instruments, and soil testing supplies, if available. Portions of 12 topographic sheets are included in the manual. Of course, you can supplement this brief presentation with your favorite topographic maps and aerial photos featuring local examples. There is also a four-color rock identification key to help students identify and classify various rock types. This key is incorporated in *Lab Exercise 21: The Rock Cycle and Rock Identification.*

This edition features links to Google Earth KMZ files for several exercises so that students can actually fly through and experience 3-D landscapes as they work problems. The exercises can be completed in a campus computer lab if Google Earth is installed or outside of class if facilities are not available on campus.

Organization: The manual begins with a unique Prologue Lab. This initial lab includes several activities that span the entire term. You may choose to assign only portions of the activities presented. These elements were designed for continuity and to help you initiate *group activities* early in the class, especially the "Hazard Identification: Natural and Anthropogenic" assignment.

The 30 exercises are subdivided into sections. Depending on your emphasis, a lab meeting may include several sections as part of an entire exercise, leaving out other sections as you see fit. We have designed the sections so that they can stand alone in most cases.

Updates: Building on the success of previous editions, this tenth edition has been extensively updated and revised. Each of the 30 lab exercises begins with a Quick Response (QR) code that students can scan on their mobile device to view a Pre-Lab Video. To view additional lab prep materials, students should visit www.masteringgeography.com. The topographic maps have all been revised for improved clarity and vibrance. Improved writing with tighter, more focused student questions and explanations are features of this edition. We continue the practice of placing some completed answers in each section to guide the student. We improved questions by requiring more interpretation of concepts under discussion. Please note new materials, work, and examples in the lab exercises on global climates, recurrence intervals for natural events, coastal geomorphology and sea-level rise, arid geomorphology, and a geographic information systems exercise with data available for download, among many others.

The features that made the previous editions such a success are retained: the *Prologue Lab* as a term-length activity drawing the students together as a continuity thread; the selection of color topographic maps and related exercises; the section structure that allows you, the teacher, to edit and assign portions within a lab and the exclusion of other sections; and other features. We updated and improved the *Answer Key* with all the work completed, illustrations drawn, calculations made, and questions answered for quick reference when grading work. The answer key can be downloaded online at www.pearsonhighered .com/irc.

Acknowledgments

We extend our continuing gratitude to the editorial, production, and sales staff of Pearson. Thanks to Christian Botting, Executive Editor of Geosciences Courseware, for willingness to risk new ideas and technologies, for his role in *Geosystems, Elemental Geosystems*, and *Geosystems Core*, and in geographic education in general. Our continuing compliments to the expertise of Emily Bornhop, Editorial Assistant, who helped oversee this Applied Physical Geography lab manual for her attention to detail and making the schedule work. We also extend our thanks to Anton Yakovlev for his years of helping to oversee the production of this lab manual and all the Geosystems projects he has helped with over the many years. Thanks to Kevin Lear and his team at International Mapping, the team at Lachina for their work on the new artwork, and Jeanine Furino for her assistance with the manuscript. And, our thanks go to the staff at Pearson for publishing what you hold in your hands.

Robert acknowledges the continuing invaluable production and photographic work of his wife Bobbé. Charlie offers thanks for the wonderful support from his wife Leslie and his children Emma and Finn in completing this work. Both of us acknowledge the role of students in our lives and the learning crucible that is the classroom.

In addition to all those geographers that contacted us by e-mail or at professional meetings, to Gail L. Hobbs and her work on several editions, and Pete Akers, Joy Mast, and Scott Mandia for their careful and thoughtful review of prior manuscripts, we thank the following colleagues for the time and effort they invested in content reviews and work on previous editions (a combined alphabetical list follows).

Maura Abrahamson, *Morton College*
Pete Akers, *University of Georgia*
Mark R. Anderson, *University of Nebraska, Lincoln*
Gregorey D. Bierly, *Indiana State University*
Mark Blumler, *SUNY, Binghamton*
Mark Bowen, *University of Wisconsin-Oshkosh*
Robin Buckallew, *Central Community College*
Peter U. Clark, *Oregon State University*
Woonsup Choi, *University of Wisconsin-Milwaukee*
Lisa DeChano-Cook, *Western Michigan University*
Carol Delong, *Victor Valley College*
John Dassinger, *Chandler-Gilbert Community College*
Shannon Fargey, *University of Victoria*
Stephen Franklin, *Coconino Community College*
Robert H. Gorcik, *William Rainey Harper College*
Tracy Galarowicz, *Central Michigan University*
Sharon Johnson, *San Francisco State University*
Chelsea Karthauser, *University of Montana*
Anil Kumar K Gangadharan, *Georgia State University*
Kory Konsoer, *Louisiana State University*
Cecil S. Keen, *Mankato State University*
James Kernan, *SUNY Geneseo*
Ingrid Luffman, *East Tennessee State University*

Scott Mandia, *SUNY Suffolk*
Joy Mast, *Carthage College*
Raoul Miller
Abdullah F. Rahman, *Indiana University*
Philip P. Reeder, *University of Nebraska, Omaha*
Betty Rhodes, *University of Wisconsin-Milwaukee*
Bradley C. Rundquist, *University of North Dakota*
Donald C. Rundquist, *University of Nebraska*
Anne Saxe, *Saddleback College*
Susan Slowly, *Blinn College*
Tak Yung Susanna Tong, *University of Cincinnati*
Eugene Turne, *California State University, Northridge*
Forrest Wilkerson, *Minnesota State University*
Thomas Williams, *Western Illinois University*
Matthew R. Zorn, *Carthage College*
Henry J. Zintambilla, *Illinois State University*

Despite this expertise and support, we accept responsibility for whatever errors persist and for those areas needing further development toward future editions. We hope that *Applied Physical Geography* teaches principles effectively and enriches your experience with the science of physical geography. This is a remarkable moment in the human—Earth experience to be enrolled in physical geography study.

Conclusion and Contact Information

We live in an extraordinary era of **Earth systems science**. This science contributes to our emerging view of Earth as a complete entity—an interacting set of physical, chemical, and biological systems that produce a whole Earth. Physical geography is at the heart of Earth systems science as we answer the spatial questions concerning Earth's physical systems and their interaction with living things. We hope that this lab is a step along the path of understanding Earth.

Contact information: As before, please consider the tenth edition of *Applied Physical Geography* an evolving work in progress. Given this admission, we welcome you to send any feedback, suggestions, criticisms, and comments to help us improve future editions. **Thank you for any assistance!**

Send comments and inquiries to:

Charles E. Thomsen
American River College
4700 College Oak Drive
Sacramento, CA 95841
E-mail: thomsec@arc.losrios.edu

Robert W. Christopherson
P. O. Box 128
Lincoln, CA 95648
E-mail: bobobbe@aol.com

The First Assignment

To begin: complete your personal Geography I.D. (see Preface, page xii). Use the maps in a physical geography text, an atlas, a college catalog, and additional library materials, if needed. Your laboratory instructor will help you find additional source materials for data pertaining to the campus.

On text or atlas maps, find the information requested, noting the January and July monthly values (the small scale of such maps will permit only a general determination); air pressures are indicated by isobars, annual temperatures indicated by the isotherms, precipitation indicated by isohyets, climatic region, landform class, soil order, and ideal terrestrial biome, and the other informa-tion requested. Record the information from source maps in the spaces provided. The completed page gives you a relevant geographic profile of your immediate environment. As you progress through your physical geography class and physical geography lab, the full meaning of these descriptions will unfold. This page might be one you want to keep for future reference.

As you contemplate further schooling in colleges and universities in other parts of the world, or perhaps when you are preparing to move to another city, you may want to prepare a Geography I.D. for these new locations as part of your investigation into future options!

GEOGRAPHY I.D.

NAME: _____ LAB SECTION: _____

HOME TOWN: _____ LATITUDE: _____ LONGITUDE: _____

COLLEGE/UNIVERSITY: _____

CITY/TOWN: _____ COUNTY (PARISH): _____

Standard Time Zone (for college location): _____

Latitude: _____ Longitude: _____

Elevation (include location of measurement on campus): _____

Place (tangible and intangible aspects that make this place unique): _____

Region (aspects of unity shared with the area; cultural, historical, economic, environmental):

_____

_____

Population: Metropolitan Statistical Area (CMSA, PMSA, if applicable): _____

Environmental Data: (information sources used: _____)

January Avg. Temperature: _____ July Avg. Temperature: _____

January Avg. Pressure (mb): _____ July Avg. Pressure: _____

Average Annual Precipitation (cm/in.): _____

Avg. Ann. Potential Evapotranspiration (if available; cm/in.): _____

Climate Region (Köppen symbol and name description): _____

_____

Main climatic influences (air mass source regions, air pressure, offshore ocean temperatures, etc.):

_____

Topographic Region or Structural Region (including rock type, loess units, etc.): _____

_____

Dominant Regional Soil Order _____

Biome (terrestrial ecosystems description; ideal and present land use): _____

_____

_____

Physical Geography Logs, Hazard I.D. Group Activity, Weather Calendar, Google Earth™

Welcome to the Prologue Lab! All of us are geographers at one time or another. Our mobility is such that we cover great distances every day. Understanding both our lives and the planet's systems requires the tools of *spatial analysis* that are at the core of geography.

We personally interact with physical systems that are expressed in complex patterns across Earth. We need to develop continuity in our observations.

Throughout this term you may be using this Prologue Lab to record your observations of various environmental phenomena. Your lab instructor will guide you as to which Prologue sections are required and the schedule for completion. These assignments are meant to span the entire term.

Key Terms and Concepts

climate
environmental events log

hazard identification: natural
   and anthropogenic
hazard perception

meteorology
weather and weather calendar

KEY**LEARNING**concepts

After completion of this lab you should be able to:

- **Analyze** Internet, printed, and published media coverage of environmental events to *discern* the spatial implications of these happenings.
- **Relate** these contemporary events to your physical geography textbook, lecture discussions, and lab, and to your daily experiences.
- **Determine** natural and anthropogenic hazards on local and regional scales.
- **Identify** and *utilize* local sources of environmental data, including broadcast and published media, instrument installations, and government sources, to *complete* a weather calendar for up to 30 days of the term, as assigned by your instructor.
- **Use** Google Earth™ to open placemarks and navigate through landscapes.

Name: _____   Laboratory Section: _____

Date: _____   Score/Grade: _____

Environmental Events Log

These are challenging times of significant global change relative to the environment, societies, and cultures, and it is hoped that this exercise will bring these important subjects to light! Studies estimate that annual weather-related damage losses (drought, floods, hail, tornadoes, derechos, tropical systems, storm surge, blizzard and ice storms, and wildfires) could exceed $1 trillion by 2040. 2016 saw 15 weather and climate events, each with losses exceeding $1 billion. These events included drought, wildfires, 4 inland floods, 8 severe storms, and a tropical cyclone. These 15 events caused 138 fatalities and cost $46.0 billion. The 2016 total was the 2nd highest annual number of U.S. billion-dollar disasters, behind the 16 events that occurred in 2011. The cost of weather-related destruction can be staggering—Hurricane Sandy in 2012 produced $65 billion in damages. In 2005, weather-related losses totaled $210 billion; Hurricane Katrina alone produced $146 billion (adjusted to current dollars) in damage that year. Weather-related destruction has risen more than 500% over the past three decades as population has increased in areas prone to violent weather and as climate change intensifies weather anomalies. The year 2016 was the warmest ever recorded, 0.99°C warmer than the 1951–1980 average, and 2016 was also the first time three consecutive warmest years have occurred since the current warming trend began in the 1970s, making it the third record-breaking year in a row. Sixteen of the 17 warmest years ever recorded have occurred since 2000. The last month that Earth saw a cooler than normal month was February 1985. 2000 to 2010 was the warmest decade ever recorded. Finally, 2016 was the fortieth consecutive year of above-average global temperatures.

Use the spaces provided to record environmental events: weather events (heavy rain, drought, winds, freezes), earthquakes, volcanic eruptions, floods and coastal inundation, tsunami events (seismic waves), biodiversity issues and species extinctions, landslides, record icebergs, Antarctic ice-shelf disintegration, stratospheric ozone updates, air pollution occurrences, or other significant events involving the physical elements of the environment. Place the number of the item in its correct location on the accompanying map(s) at the end of this log, perhaps color-coding the number according to the type of event (blue for weather, brown for geomorphology, green for biogeography, etc.). A classroom bulletin board map might be used to compile a class world/North American map of environmental events.

Please access available media (Internet, broadcast, newspapers and magazines, and journals) as time permits. Use public television (PBS) and National Public Radio (NPR) shows such as *NOVA, Lehrer NewsHour, National Geographic, Nature,* and National Public Radio's *Morning Edition* and *All Things Considered,* or any other sources. *Geosystems,* tenth edition, lists more than 200 URLs for Internet sources. The Internet is particularly rich for learning about the latest occurrences.

You will see how much of physical geography is actually occurring out there in "the real world." Your list is not meant to be comprehensive; rather, it should reflect a growing interest in the physical systems that surround our lives. Restrict your list to events that occur throughout the term or semester. Spaces are provided for 18 items.

1. Date   **Source**   **Event**

_____; _____;

Tie-in to physical geography:

    2. Date   **Source**   **Event**

_____; _____;

Tie-in to physical geography:

    3. Date   **Source**   **Event**

_____; _____;

Tie-in to physical geography:

    4. Date   **Source**   **Event**

_____; _____;

Tie-in to physical geography:

    5. Date   **Source**   **Event**

_____; _____;

Tie-in to physical geography:

    6. Date   **Source**   **Event**

_____; _____;

Tie-in to physical geography:

    7. Date   **Source**   **Event**

_____; _____;

Tie-in to physical geography:

8. Date   **Source**   **Event**

_____; _____;

Tie-in to physical geography:

9. Date   **Source**   **Event**

_____; _____;

Tie-in to physical geography:

10. Date   **Source**   **Event**

_____; _____;

Tie-in to physical geography:

11. Date   **Source**   **Event**

_____; _____;

Tie-in to physical geography:

12. Date   **Source**   **Event**

_____; _____;

Tie-in to physical geography:

13. Date   **Source**   **Event**

_____; _____;

Tie-in to physical geography:

14. Date   **Source**   **Event**

_____; _____;

Tie-in to physical geography:

15. Date   **Source**   **Event**

_____; _____;

Tie-in to physical geography:

16. Date   **Source**   **Event**

_____; _____;

Tie-in to physical geography:

17. Date   **Source**   **Event**

_____; _____;

Tie-in to physical geography:

18. Date   **Source**   **Event**

_____; _____;

Tie-in to physical geography:

▲ **Figure 1** Locations of environmental events in the world (locate and label)

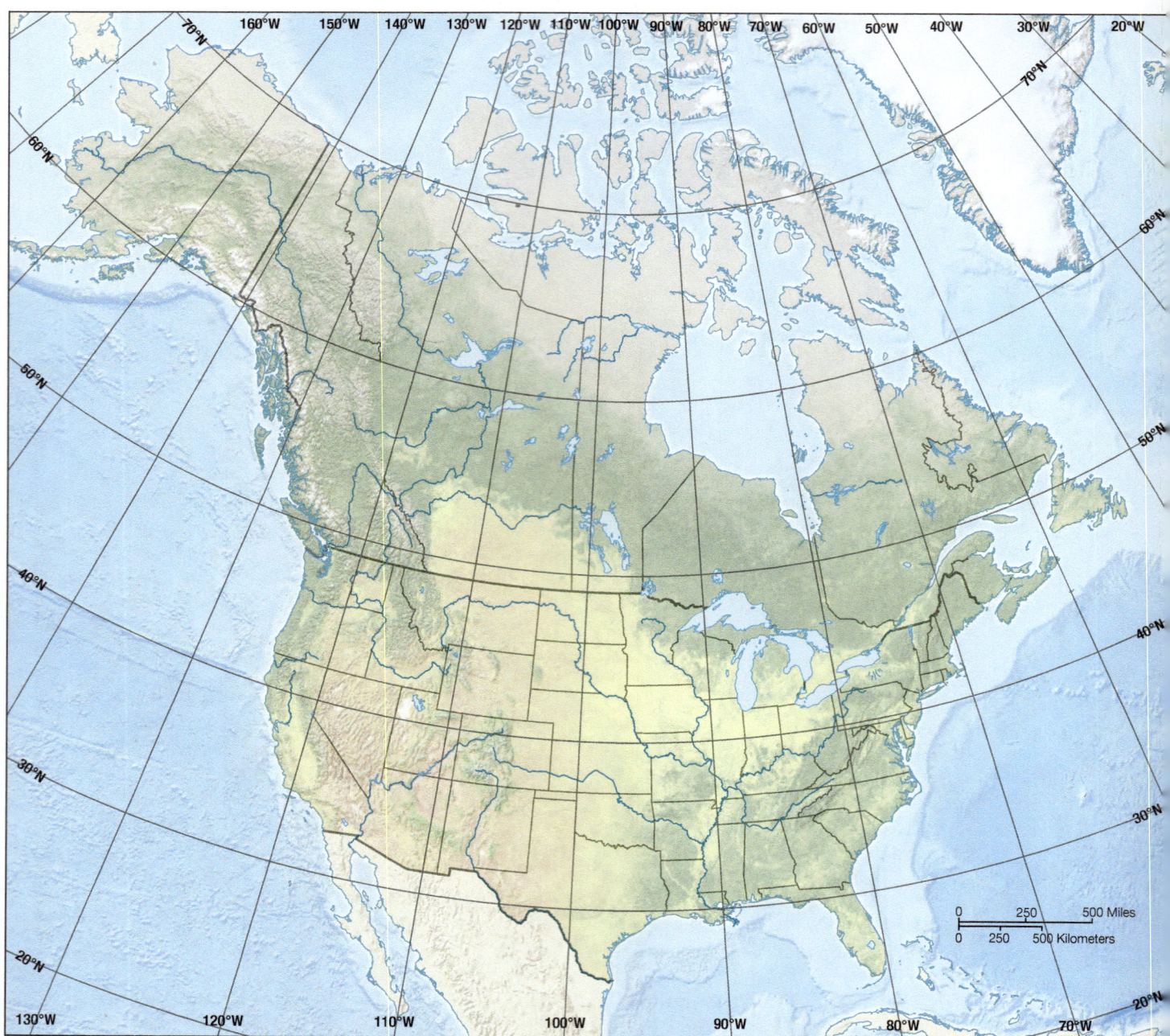

▲ **Figure 2** Locations of environmental events in the United States and Canada (locate and label)

Name: _____   Laboratory Section: _____

Date: _____   Score/Grade: _____

Hazard Identification

Natural Hazards and Anthropogenic (Human-Forced) Hazards—A Group Activity—

A valid and applicable generalization is that humans are unable or unwilling or incapable of perceiving hazards in a familiar environment. Such an axiom of human behavior certainly helps explain why large populations continue to live and work in hazard-prone settings. Similar questions also can be raised about populations in areas vulnerable to natural disasters. Political and economic interests ride this wave of poor public perception and lack of physical geography's spatial perspective. Ideally, an informed public that acts on its knowledge can move to prevent development in hazardous areas. This poor perception applies to human-induced hazards as well, such as groundwater contamination, air pollution, and radiological hazards.

A consequence of poor hazard perception is that our entire society bears the financial cost, the subsequent bad planning, and irrational development—no matter where in the country it occurs. Of course, those directly affected are the victims who must shoulder the physical, emotional, and economic hardships of the disaster and for which society indirectly pays. This recurrent yet avoidable cycle—construction, devastation, reconstruction, devastation—continues. Recent events are reminders of this problem in our relationship with nature. A few of the many examples include the record-setting 2010 Northern Hemisphere heat waves that caused almost 18 million deaths, the Australian heat wave of 2016 that required a new color for their temperature maps; the devastating

tornadoes of the past decades, with more than 1600 hitting the United States alone in 2004; the Midwest floods of 2011, 2013, and 2016, the Colorado floods of 2013, and the European floods of 2013 and 2014; the nor'easters that struck the east coast during January 1998 and January 2005; Hurricanes Katrina, Rita, and Wilma in 2005; Cyclone Sidr in 2007, Hurricane Sandy in 2012, and the record-breaking Hurricane Patricia in 2015; the October 2005 earthquake in Pakistan that killed 80,000, the China earthquake of 2008 that killed 87,000, the Haiti earthquake of 2010 that killed more than 316,000, the more than 225,000 killed by the tsunamis caused by the magnitude 9.0 earthquake near the northwest coast of Sumatra, Indonesia in 2004, the Nepal and Chile earthquakes of 2015; the 2010 eruption of Mount Merapi that displaced 350,000 people and the 2010 eruption of Eyjafjallajökull that disrupted air travel for months; and the devastating 2011 9.0 earthquake off of Japan, followed by the tsunami and the nuclear power plant meltdowns. The increasing pace of global climate change demands that we pay attention to our relationships with Earth systems.

A value in studying physical geography is a better knowledge of the environment and the workings of natural systems. This heightened awareness may lead to a more reasonable assessment of natural and anthropogenic hazards. Following group discussions with lab partners, describe and explain any natural or human-induced hazards that exist within these radii from campus. Note how they affect you.

10 km (6 mi) radius of your campus:

Natural– _____

_____

_____

_____

_____

Anthropogenic– _____

_____

_____

_____

_____

100 km (60 mi) radius of your campus:

Natural– _____

_____

_____

_____

_____

Anthropogenic– _____

_____

_____

_____

_____

Your state or province:

Natural– _____

_____

_____

_____

_____

Anthropogenic– _____

_____

_____

_____

_____

Additional specific *regional* environmental issues:

_____

_____

_____

_____

Name: _____   Laboratory Section: _____

Date: _____   Score/Grade: _____

Weather Calendar

Weather is an important integrative subject within physical geography. Temperature, air pressure, relative humidity, wind speed and direction, day length, and Sun angle are important measurable elements that contribute to the weather. **Weather** is the short-term condition of the atmosphere, as compared to **climate**, which reflects long-term atmospheric conditions and extremes.

We go online or tune to a local station for the day's weather report from the National Weather Service (in the United States, **http://www.weather.gov/**) or the Canadian Meteorological Center (in Canada, **http://weather.gc.ca/index_e.html**) to see the current satellite images and to hear tomorrow's forecast. Most cable systems carry the *Weather Channel*, which presents local weather information throughout the day (**http://www.weather.com**). Internationally, the World Meteorological Organization coordinates weather information (see **http://www.wmo.int/pages/index_en.html**).

Although this is early in the term and weather has not specifically been discussed in lectures, you will find this exercise an interesting, ongoing work-in-progress that will stimulate many questions and prepare you for those chapters when it is covered.

Meteorology is the scientific study of the atmosphere (*meteor* means "heavenly" or "of the atmosphere"). Embodied within this science is a study of the atmosphere's physical characteristics and motions; related chemical, physical, and geologic processes; the complex linkages of atmospheric systems; and weather forecasting. Computers permit the handling of enormous amounts of data on water vapor, clouds, precipitation, and radiation for accurate forecasting of near-term weather.

New developments in supercomputers and ground- and space-based observation systems are making this a time of dramatic developments. A few of the innovations include wind profilers that use radar to profile winds from the surface to high altitudes, computer-based modeling software for 3-D weather models and displays, improved international cooperation and weather data dissemination, and the standard Advanced Weather Interactive Processing System (AWIPS) at more than 150 stations. Data gathering is enhanced by the completed installation of 883 Automated Surface Observing System (ASOS) instrument arrays as a primary surface weather-observing network. By the end of 2013, 160 WSR-88D (Weather Surveillance Radar) Doppler radar systems as part of the NEXRAD (Next-Generation Radar) program were operational through the National Weather Service, in conjunction with the Federal Aviation Administration and the Department of Defense. In Canada, the National Radar Project will have 30 CWSR-98 radars in service.

NOAA operates the Earth System Research Laboratory, Boulder, Colorado, which is developing new forecasting tools (see **http://www.esrl.noaa.gov**). Research and monitoring of violent weather is centered at NOAA's National Severe Storms Laboratory in several cities (see **http://www.nssl.noaa.gov**). For tropical weather, see the National Hurricane Center, at **http://www.nhc.noaa.gov**.

Building a database of atmospheric conditions is key to *numerical weather prediction* and the development of weather forecasting models. *Synoptic analysis* involves the characterization of weather elements over a region within the same time frame. You will not be observing all these elements for your weather calendar, but a complete weather forecast is based on knowing the following nine items (the seven marked with asterisks are required for this exercise):

- Barometric pressure*
- Pressure tendency
- Surface air temperature*
- Dew-point temperature*
- Wind speed and direction*
- Type of clouds*
- Current weather*
- State of the sky*
- Precipitation since last observation

The purpose of this weather exercise is twofold. *First*, it will help you find sources of weather information. Our lives take place within an environment of variable weather, so it is important to have access to such information.

Second, you will gather data for up to 30 days (or less, as assigned by your instructor) during the term, not necessarily consecutive days, and record them on a weather calendar using official international symbols and designations. If your lab meets 15 or 16 times during the term and the department has weather instruments, then half your days of observation are easily completed on class days. The

other days can be met by outside sources or visits to the department.

The National Weather Service and the Canadian Meteorological Center use a standardized system of recording and reporting weather data around a weather station symbol developed by the World Meteorological Organization (WMO). An example of this method is shown below.

WEATHER STATION SYMBOL

Attached is your "Weather Calendar," with spaces for up to 30 days of weather observations. These days need not be consecutive and may be spread out over the entire term. The completed calendar is due at the end of the term. Your instructor will give you more details. The circle in the middle of each square is the weather station symbol, and above and below are spaces to list the date and the place of observation. (If you are using a broadcast source, list the place the weather instruments are located that made the observation.)

We can refer to the weather station symbol as if it were a clock. Temperature is recorded in the 10:00 position, barometric pressure at 2:00, weather type at 9:00, and dew-point temperature (if available) at 8:00; wind speed and direction vary in location, depending on the wind itself. If you are observing clouds, high clouds are symbolized above the station symbol, and middle and low clouds below it.

The state of the sky (cloud cover) is the area inside the circle. In the example to the left, temperature is 74°F (23°C), pressure is 1004.0 mb, the sky is 50% (five-tenths) overcast, with slight continuous fall of snowflakes at time of observation. Winds are from the northwest (line drawn on the side of the station where the winds originate) at 15–20 mph (24–32 kmph), as noted by the "flags" on the line.

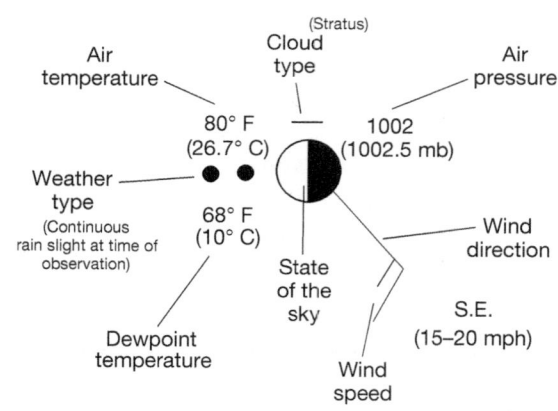

Name: _____   Laboratory Section: _____

Weather Calendar

Date/time_____ ◯ Place_____	Date/time_____ ◯ Place_____	Date/time_____ ◯ Place_____	Date/time_____ ◯ Place_____	Date/time_____ ◯ Place_____
Date/time_____ ◯ Place_____	Date/time_____ ◯ Place_____	Date/time_____ ◯ Place_____	Date/time_____ ◯ Place_____	Date/time_____ ◯ Place_____
Date/time_____ ◯ Place_____	Date/time_____ ◯ Place_____	Date/time_____ ◯ Place_____	Date/time_____ ◯ Place_____	Date/time_____ ◯ Place_____
Date/time_____ ◯ Place_____	Date/time_____ ◯ Place_____	Date/time_____ ◯ Place_____	Date/time_____ ◯ Place_____	Date/time_____ ◯ Place_____
Date/time_____ ◯ Place_____	Date/time_____ ◯ Place_____	Date/time_____ ◯ Place_____	Date/time_____ ◯ Place_____	Date/time_____ ◯ Place_____
Date/time_____ ◯ Place_____	Date/time_____ ◯ Place_____	Date/time_____ ◯ Place_____	Date/time_____ ◯ Place_____	Date/time_____ ◯ Place_____

Google Earth™

Virtual Geography in *Applied Physical Geography*

Incorporated in this edition of the Lab Manual are exercises built around Google Earth™. This technology is the most exciting and useful geographic tool since GPS (Global Positioning System) technology became available to the public. This program allows you to view a photo-realistic 3-D model of Earth. You can zoom in and out, tilt and rotate the view, adjust the vertical exaggeration of the landscape, and add your own image overlays. Some of the exercises in the Lab Manual involve simply navigating around terrain features to gain a better understanding of how they look, while other exercises involve the application of critical thinking skills.

Google Earth™ is now available as a web page at **https://google.com/earth**. While the web page allows you to add KMZ placemark files, for more control you may wish to download the desktop version of the program. The Google Earth™ website has information on the hardware and network requirements for the program. It also has an excellent tour of the features of the program and a gallery of downloadable placemarks called KMZ or KML files.

Once you have installed it, you should go through Google's tour of the program. The tour covers the basics of how to navigate around the virtual world, how to search for places, and how to open and save placemarks.

The exercises in this manual all involve downloading KMZ files from the Geosystems website and opening them in Google Earth™. This introductory exercise will help acquaint you with the process of opening a KMZ file and using it with the questions of a lab exercise. The first KMZ file is called intro. kmz. For the KMZ file and the exercise questions, go to **http://mygeoscienceplace.com** or the Study Area of Mastering Geography. This web page will have the KMZ files and the questions for each assignment. Depending on how your web browser is set up, the KMZ file will automatically open up in Google Earth™ or you may have to download it to your local machine and open it up manually.

Once you have opened the KMZ, it will appear in the sidebar in the Places box, in the Temporary Places folder. When you double-click on the Intro placemark, it will fly you to the first location and open up a pop-up window with information about that place. Some of the placemarks will have only one location; others may have several locations that you will visit, as well as overlay images that you can turn on and off. You can control the transparency of an image overlay with the transparency slider control. This will be useful when you want to interpret information from multiple overlays. After you have found your house (which is the first thing almost everyone does), go through each of the questions in the Intro assignment. The questions for each assignment are found on the same website as the KMZ files. Your instructor will tell you how to submit your assignments.

Mastering Geography

Looking for additional lab prep and review materials? Visit the Study Area in Mastering Geography to enhance your geographic literacy and spatial reasoning skills by accessing MapMaster™ interactive maps, videos, *Mobile Field Trips*, *Project Condor* Quadcopter Videos, *In the News* articles, flashcards, web links, self-study quizzes, and other resources. Contact your professor to find out how you can access Mastering Geography.

Name: _____   **Laboratory Section:** _____

Date: _____   **Score/Grade:** _____

Video
Exercise 1
Pre-Lab Video

http://goo.gl/pQk8vC

**Scan to view the
Pre-Lab video**

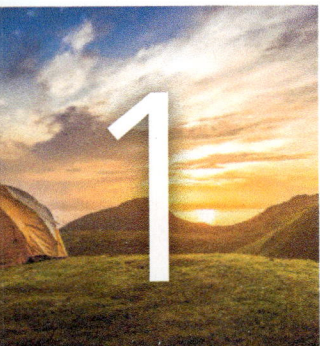

LAB EXERCISE

Latitude and Longitude

To know specifically where something is located on Earth's surface, a coordinated grid system is needed, one that is agreed to by all peoples. The terms *latitude* and *longitude* were in use on maps as early as the first century C.E., with the concepts themselves dating back to Eratosthenes and others.

As you complete your *Geography I.D.* in the Preface, you will eventually determine the latitude and longitude for your campus. We examine great circles, latitude, and longitude in this exercise. Lab Exercise 1 features three sections.

Key Terms and Concepts

great circle

latitude

longitude or arc distance

meridian

parallel

prime meridian

small circle

KEY**LEARNING**concepts

After completion of this lab you should be able to:

- **Define** great circle and small circle and *describe* the relationship between a great circle and travel.
- **Define** latitude and parallel, longitude and meridian, and *use* them in simple sketches to *demonstrate* how Earth's reference grid is established.

Materials/Sources Needed

globe

string

ruler

protractor

world atlas

Lab Exercise and Activities

SECTION 1

Great and Small Circles

Great circles and small circles are important concepts relating to latitude and longitude—concepts that we will refer to in later activities. A **great circle** is any circle of Earth's circumference whose center coincides with the center of Earth. An infinite number of great circles can be drawn on Earth. On most flat maps, airline and shipping routes appear to arch their way across oceans and landmasses following great circles. Despite their curved appearance on flat maps, great circle routes are the shortest distance between two points on Earth.

Using a globe and a piece of string, place one end on San Francisco and stretch the string tautly to London. This direct route is a portion of a great circle. **Small circles** are circles whose centers do not coincide with Earth's center. In **Figure 1.1** properly draw and label a great circle and a small circle, other than those noted by the latitude and longitude grid.

▲ **Figure 1.1** Label great circles and small circles

Again using the globe and piece of string, show the great circle routes linking the following cities, listing at least four or five prominent geographical features (e.g., mountains, seas, rivers, nations, cities, etc.) that you would cross as you traveled along those routes:

1. London, England, and Colombo, Sri Lanka (English Channel, Germany, Carpathian Mtns.)

2. Vancouver, British Columbia, and Sydney, Australia

3. Your hometown and Beijing, China

SECTION 3

Temperature Readings

Relative to atmospheric temperature, locate a properly installed thermometer either at the college or university you are attending or perhaps at home. If you do not have access to a thermometer, find another source of information for local temperature: a radio or television station, a local cable channel, the Weather Channel, a local newspaper, the National Weather Service, Environment Canada, or one of the several Internet sources available.

Perhaps you have already done this in working on your "Weather Calendar" in the Prologue Lab. The purpose of this section is to identify a reliable source of temperature information. Long after this lab course is completed, such information will benefit your lifestyle and activities. Also, this information is important to energy management of interior spaces.

1. Record the air temperature for 4 days, at approximately the same time of day, if possible. See if you can detect a trend between air temperature and other atmospheric phenomena.

Temperature	Place of Observation	Time of Day; Date	Weather conditions	State of the sky
Day 1:	Place:	Time:		
Day 2:	Place:	Time:		
Day 3:	Place:	Time:		
Day 4:	Place:	Time:		

2. Do you have an outdoor thermometer available where you live? Describe the location of the thermometer from which you obtained these readings (orientation, exposure relative to compass direction, height from the ground).

3. Is temperature data available through the Geography Department at your school or elsewhere on campus? Is there a campus weather station?

Mastering Geography

Looking for additional lab prep and review materials? Visit the Study Area in Mastering Geography to enhance your geographic literacy and spatial reasoning skills by accessing MapMaster™ interactive maps, videos, *Mobile Field Trips, Project Condor* Quadcopter Videos, *In the News* articles, flashcards, web links, self-study quizzes, and other resources. Contact your professor to find out how you can access Mastering Geography.

NOTES

LAB EXERCISE

Temperature Patterns

Principal controls and influences on temperature patterns include latitude, elevation, cloud cover, land–water heating differences, ocean currents and sea-surface temperatures, and general surface conditions. A variety of temperature regimes worldwide affect cultures, decision making, and resource consumption. Global temperature patterns apparently are changing in a warming trend and are the subject of much scientific, geographic, and political interest. In Lab Exercise 8 we analyze the spatial patterns of global temperatures. Lab Exercise 8 features two sections.

Key Terms and Concepts

average annual temperature
continentality
land–water heating differences

marine (maritime)
normal lapse rate

specific heat
temperature range

KEY**LEARNING**concepts

After completion of this lab, you should be able to:

- **Contrast** and *compare* the temperature patterns of La Paz and Concepción, Bolivia, and *determine* the effects of elevation on temperatures.
- **Compare** and *contrast* the temperature patterns of Vancouver, British Columbia, and Winnipeg, Manitoba, and *determine* the differing effects of marine and continental factors.
- **Plot** and *graph* temperature data on climographs for selected stations.

Materials/Sources Needed

pencil
colored pencils

highlighter pen
calculator

Lab Exercise and Activities

SECTION 1

Elevation and Temperature

Within the troposphere, temperatures decrease with increasing altitude above Earth's surface: The **normal lapse rate** of temperature change with altitude is 6.4 C°/1000 m, or 3.5 F°/1000 ft. Worldwide, mountainous areas experience lower temperatures than do regions nearer sea level, even at similar latitudes. The consequences are that average air temperatures at higher elevations are lower, nighttime cooling increases, and the temperature range between day and night and between areas of sunlight and shadow also increases. Temperatures may decrease noticeably in the shadows and shortly after sunset. Surfaces both gain heat rapidly and lose heat rapidly to the thinner atmosphere.

Temperature data are presented in **Table 8.1** for La Paz and Concepción. Both stations are approximately the same distance south of the equator but differ in elevation. La Paz is at 4103 m (13,461 ft),
whereas Concepción is at 490 m (1608 ft) above sea level. The hot, humid climate of Concepción at its much lower elevation stands in marked contrast to the cool, dry climate of highland La Paz.

People living around high-elevation La Paz actually grow wheat, barley, and potatoes—crops characteristically grown in the cooler midlatitudes at lower elevations. These crops do well despite the fact that La Paz is 4103 m above sea level. The combination of *elevation* (moderating temperatures) and *equatorial location* (producing higher Sun altitude and consistent daylength) guarantee La Paz these temperature conditions, averaging about 9°C (48°F) for every month. Such moderate temperature and moisture conditions lead to the formation of more fertile soils than those found in the warmer, wetter climate of Concepción.

1. Using the temperature graphs provided in **Figure 8.1**, plot the data from Table 8.1 for these two cities. Use a smooth curved *line graph* to portray the temperature data. Calculate the **average annual temperature** and the **temperature range** (difference between the highest and lowest) for each city.

2. Why are the temperatures at La Paz more moderate in every month and so consistent overall as compared to Concepción?

3. Recall that the *normal lapse rate* of temperature change with altitude is 6.4 C°/1000 m, or 3.5 F°/1000 ft. Calculate the difference in elevation between La Paz and Concepción and calculate what the difference in their *mean (average) annual temperatures* should be, based on cooling at the normal lapse rate. Is the actual difference between the mean annual temperatures for these two cities higher, lower, or the same as that produced by calculating average normal lapse rate conditions? (Show your work.)

4. The annual march of the seasons and the passage of the subsolar point between the Tropics of Cancer and Capricorn affect these stations. Can you detect from your temperature graphs these seasonal effects? Explain.

La Paz, Bolivia

Avg. Annual Temperature (°C, °F): _____

Annual Temp. Range (C°, F°): _____

Concepción, Bolivia

Avg. Annual Temperature (°C, °F): _____

Annual Temp. Range (C°, F°): _____

▲ **Figure 8.1** Temperature graphs for La Paz and Concepción, Bolivia

TABLE 8.1 Temperature data for La Paz and Concepción, Bolivia												
La Paz, Bolivia: pop. 877,363, lat. 16.5°S, long. 68.17°W, elev. 4103 m (13,461 ft).												
	Jan	Feb	Mar	Apr	May	Jun	Jul	Aug	Sep	Oct	Nov	Dec
Avg. Temp.°C	9.0	9.0	9.0	9.0	8.0	7.0	8.0	8.0	9.0	10.0	10.0	10.0
(°F)	(48.2)	(48.2)	(48.2)	(48.2)	(46.4)	(44.6)	(46.4)	(46.4)	(48.2)	(50.0)	(50.0)	(50.0)
Concepción, Bolivia: pop. 8,221, lat. 16.25°S, long. 62.05°W, elev. 490 m (1608 ft).												
	Jan	Feb	Mar	Apr	May	Jun	Jul	Aug	Sep	Oct	Nov	Dec
Avg. Temp.°C	25.0	26.0	25.0	24.0	23.0	19.0	22.0	24.0	27.0	27.0	26.0	26.0
(°F)	(77.0)	(78.8)	(77.0)	(75.2)	(73.4)	(66.2)	(71.6)	(75.2)	(80.6)	(80.6)	(78.8)	(78.8)

SECTION 2

Marine vs. Continental Effects

The irregular arrangement of landmasses and water bodies on Earth contributes to the overall pattern of temperature. The physical nature of the substances themselves—rock and soil versus water—is the reason for these **land–water heating differences**. More moderate temperature patterns are associated with water bodies compared to more extreme temperatures inland.

These contrasts in temperatures are the result of the land–water temperature controls: evaporation, transmissibility, **specific heat**, movement, ocean currents, and sea-surface temperatures. The term **marine**, or *maritime*, is used to describe locations that exhibit the moderating influences of the ocean, usually along coastlines or on islands. **Continentality** refers to the condition of areas that are less affected by the sea and therefore have a greater range between maximum and minimum temperatures diurnally (daily) and yearly.

The Eurasian cities of Trondheim, Norway, Ukhta, Russia, and Yakutsk, Russia, exemplify these marine and continental conditions. All three cities are at approximately 63°N. They are between 11 m (36 ft) and 99 m (325 ft) elevation. However, Trondheim has a more moderate pattern of average maximum and minimum temperatures than do either of the other two. Trondheim's annual range of 16.9 C° (30.4 F°) is far below the 33.0 C° (59.4 F°) temperature range of Ukhta and the extreme range of Yatusk, 58.1 C° (104.5 F°). In fact, Yakutsk's continental temperature pattern is more extreme in every aspect than that of maritime Trondheim.

1. Using the data given in **Table 8.2**, plot the temperatures for these three cities and portray with a smooth curved *line graph* on the temperature graph in **Figure 8.3**. Calculate the average annual temperature and temperature range for each city.

▲ **Figure 8.2** Location map of Trondheim, Norway, Ukhta, Russia, and Yakutsk, Russia

Trondheim, Norway

Avg. Annual Temperature (°C, °F): _____

Annual Temp. Range (C°, F°): _____

Ukhta, Russia

Avg. Annual Temperature (°C, °F): _____

Annual Temp. Range (C°, F°): _____

▲ **Figure 8.3** Temperature graphs for Trondheim, Norway, Ukhta, Russia, and Yakutsk, Russia.

TABLE 8.2 Temperature data for Trondheim, Norway, Ukhta, Russia, and Yakutsk, Russia.											
Trondheim, Norway: pop. 187,353, lat. 63°25′, long. 10°23′, elev. 11 m											
	Jan	Feb	Mar	Apr	May	Jun	Jul	Aug	Sep	Oct	Nov	Dec
Temp.°C	−1.9	−1.6	0.9	4.9	9.2	12.5	15	14.2	10.6	6.1	1.4	−1.3
(°F)	(28.6)	(29.1)	(33.6)	(40.8)	(48.6)	(54.5)	(59)	(57.6)	(51.1)	(43)	(34.5)	(29.7)
Ukhta, Russia: pop. 99,951, lat. 63°34′N, long. 53°42′E, elev. 82 m											
	Jan	Feb	Mar	Apr	May	Jun	Jul	Aug	Sep	Oct	Nov	Dec
Temp.°C	−16.5	−14.7	−6.8	−0.5	6.3	13.3	16.5	12.4	6.9	0.4	−8.9	−13.8
(°F)	(2.3)	(5.5)	(19.8)	(31.1)	(43.3)	(55.9)	(61.7)	(54.3)	(44.4)	(32.7)	(16)	(7.2)

2. Compare and contrast the marine temperature regimes of Trondheim with the continental regimes of Ukhta and Yakutsk. What significant differences do you note?

Yakutsk, Russia: pop. 269,601, lat. 63°02', long. 129°44', elev. 99 m											
	Jan	Feb	Mar	Apr	May	Jun	Jul	Aug	Sep	Oct	Nov	Dec
Temp. °C	−38.6	−33.6	−20.1	−4.8	7.5	16.4	19.5	15.2	6.1	−7.8	−27	−37.6
(°F)	(−37.5)	(−28.5)	(−4.2)	(23.4)	(45.5)	(61.5)	(67.1)	(59.4)	(43)	(18)	(−17)	(−35.7)

Yakutsk, Russia

Avg. Annual Temperature (°C, °F): _____

Annual Temp. Range (C°, F°): _____

Months

3. How many months register average temperatures below freezing in Trondheim, Ukhta, and Yakutsk? How many months above 10°C does each city have? Explain.

Mastering Geography

Looking for additional lab prep and review materials? Visit the Study Area in Mastering Geography to enhance your geographic literacy and spatial reasoning skills by accessing MapMaster™ interactive maps, videos, *Mobile Field Trips*, *Project Condor* Quadcopter Videos, *In the News* articles, flashcards, web links, self-study quizzes, and other resources. Contact your professor to find out how you can access Mastering Geography.

Name: _____   Laboratory Section: _____

Date: _____   Score/Grade: _____

LAB EXERCISE

Temperature Maps

Principal controls and influences on temperature patterns include latitude, elevation, cloud cover, land–water heating differences, ocean currents and sea-surface temperatures, and general surface conditions. A variety of temperature regimes worldwide affect cultures, decision making, and resource consumption. Global temperature patterns apparently are changing in a warming trend and are the subject of much scientific, geographic, and political interest. In Lab Exercise 9, we analyze the spatial patterns of global temperatures. Lab Exercise 9 features two sections.

Key Terms and Concepts

isotherm   thermal equator

KEY**LEARNING**concepts

After completion of this lab, you should be able to:

- **Construct** isotherms on a U.S.–Canadian map and *analyze* resultant temperature patterns portrayed.
- **Analyze** global temperature patterns by plotting temperature profiles.

Materials/Sources Needed

pencil   highlighter pen
colored pencils   calculator

SECTION 1

Working with Isotherms and Temperature Maps

Both the pattern and variety of temperature on Earth are outputs of the global energy system. Temperature maps portray the interaction of the principal temperature controls: latitude, altitude, cloud cover, and land–water heating differences. To depict the spatial aspect of data, geographers utilize *isopleths* (also called *isolines*)—lines drawn on a map connecting points of equal value, such as contour lines on a topographic map that show points of equal elevation (Lab Exercise 23).

An isopleth on temperature maps is known as an **isotherm** (similarly, pressure maps = isobars, precipitation maps = isohyets, and equal-wind velocity = isotachs). An isotherm connects points of equal temperature and presents the pattern of temperature for analysis. Geographers are concerned with the spatial distribution of temperatures, and isotherms facilitate this analysis.

In Lab Exercise 23, Section 2, you will construct contour lines to show elevation change. Isotherms, connecting equal temperature values, are drawn in much the same way. Construct the appropriate isotherms at 5 C° intervals on the weather map in **Figure 9.1**. The surface weather map data were recorded at 7:00 A.M. EST on October 30, 1993.

1. The 0°C (32°F) isotherm is already drawn for you (highlight it to make it distinctive). You may find it helpful to use colored pencils to "group" the temperature intervals (5 C° intervals, above and below 0°C). Use a pencil to lightly sketch the isotherms, then use a pen to darken the lines, which should be smooth curved lines. (See your geography text or the weather page of the newspaper for examples of isothermal maps.) Begin with the 5°C isotherm, then draw the 10°C and 15°C isotherms. Continue by completing the +5°C, +10°C, +15°C, +20°C, and finally, +25°C isotherms.

2. List and briefly describe the principal temperature controls on average annual temperatures and temperature ranges.

3. Describe the temperature pattern from Louisiana northward through Arkansas, Missouri, Iowa, and Wisconsin, to Canada. Which temperature controls are producing this pattern?

4. Describe and explain the temperature gradient over Nevada and Utah. (How rapidly are the temperatures changing from one place to another?) Which temperature controls are producing this pattern of temperatures.

5. What is the temperature range on this map? Calculate the difference between the maximum (highest) and minimum (lowest) temperature readings. In the spaces below, fill in the readings; use a map or an atlas to determine the city (station) where the data were collected.

   Maximum temperature: _____ Station (city) _____

   Minimum temperature: _____ Station (city) _____

   Range (express in C° and F°): _____

Temperatures in °C (°C times 1.8 + 32 = °F)
Isotherm interval 5 C°

------- 0° (32°F)

Daily Weather Map published by the Meteorological Operations Division and Climate Analysis, National Meteorological Center, National Weather Service, National Oceanic and Atmospheric Administration, Department of Commerce, 1993.

Temperature data derived from the *Daily Weather Map* for October 30, 1993, 7:00 A.M. E.S.T.

▲ **Figure 9.1** Isotherm map

Global Surface Temperature Map Analysis

Just as closer contour lines on a topographic map (Lab Exercise 23) mark a steeper slope, so do closer isotherms denote steepness in the temperature gradient (change across Earth's surface). In **Figure 9.2a** (January), note the spacing of the isotherms along the 90°W meridian between 40° and 70°N latitudes (steep temperature gradient) and between 40° and 70°S latitudes (gradual temperature gradient). The **thermal equator**, the location where the highest mean temperature is found is shown by a dashed line. Isotherms spaced at greater distances from one another mark a more gradual temperature change. Relate these patterns to the distribution of land and water surfaces—water is generally more moderate, land more extreme, in temperature distributions.

As you use the maps in Figures 9.2a and **9.2b**, remember that these maps use a 3°C interval (example: 0°C, 3°C, 6°C, 9°C, etc.). If the data points you are plotting on the charts in **Figures 9.3a**, **9.3b**, **9.4a**, and **9.4b** fall between two isotherms, you can interpolate (estimate) the temperature value to mark on the graph.

1. Temperature profiles, like topographic profiles (Lab Exercise 23), illustrate how rapidly temperatures change over Earth's surface. Use the four graphs in Figures 9.3a, 9.3b, 9.4a, and 9.4b to plot temperature data along two parallels and two meridians as noted; then complete the temperature profiles with a line graph connecting the plotted data points:

    a) January (Figure 9.2a), along 50°N parallel (already plotted and drawn for you)

    b) January (Figure 9.2a), along 90°W meridian (already plotted and drawn for you)

    c) July (Figure 9.2b), along 40°N parallel

    d) July (Figure 9.2b), along 60°E meridian

    After completing the two graphs in Figures 9.4a and 9.4b, shade the parts of your plots that are over land in brown and over water in blue along the bottom row or two of boxes or the left column or two of boxes (done for you in Figure 9.3b), then complete the following.

2. Describe the pattern of average temperatures as you travel from west to east at 50°N in January and 40°N in July. Where are temperatures higher/lower? Contrast temperatures over the land and over the oceans.

    a) January:

    b) July:

3. Using these maps and graphs, describe and explain the patterns of temperature gradients on Figures 9.4a and 9.4b: Where are gradients steep or gentle?

    a) January:

    b) July:

4. Describe and explain the differences in temperature patterns between the Northern and Southern Hemispheres in January and July. What is the general pattern of isotherms? What are the extreme highs and lows in each hemisphere in January and July? Which hemisphere has the least amount of seasonal change?

▲ Figure 9.2a Global temperatures for January (°C)

▲ **Figure 9.2b** Global temperatures for July (°C)

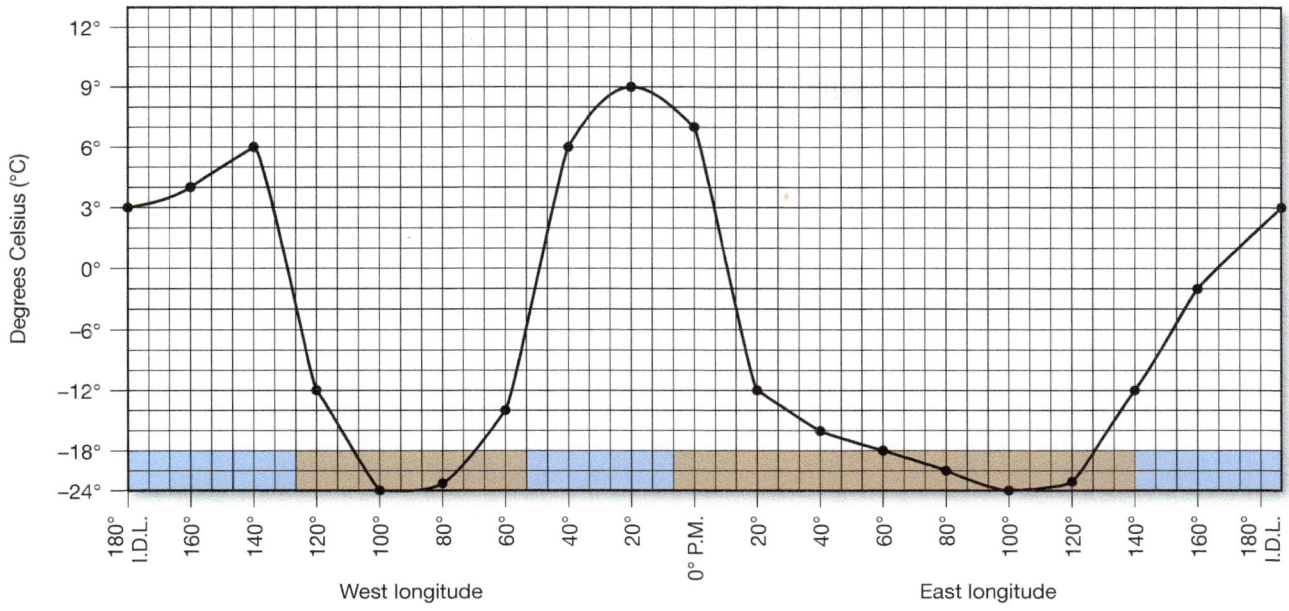

▲ **Figure 9.3a** Plot of average temperatures along 50°N latitude for January

▲ **Figure 9.3b** Plot of average temperatures along 90°W longitude for January

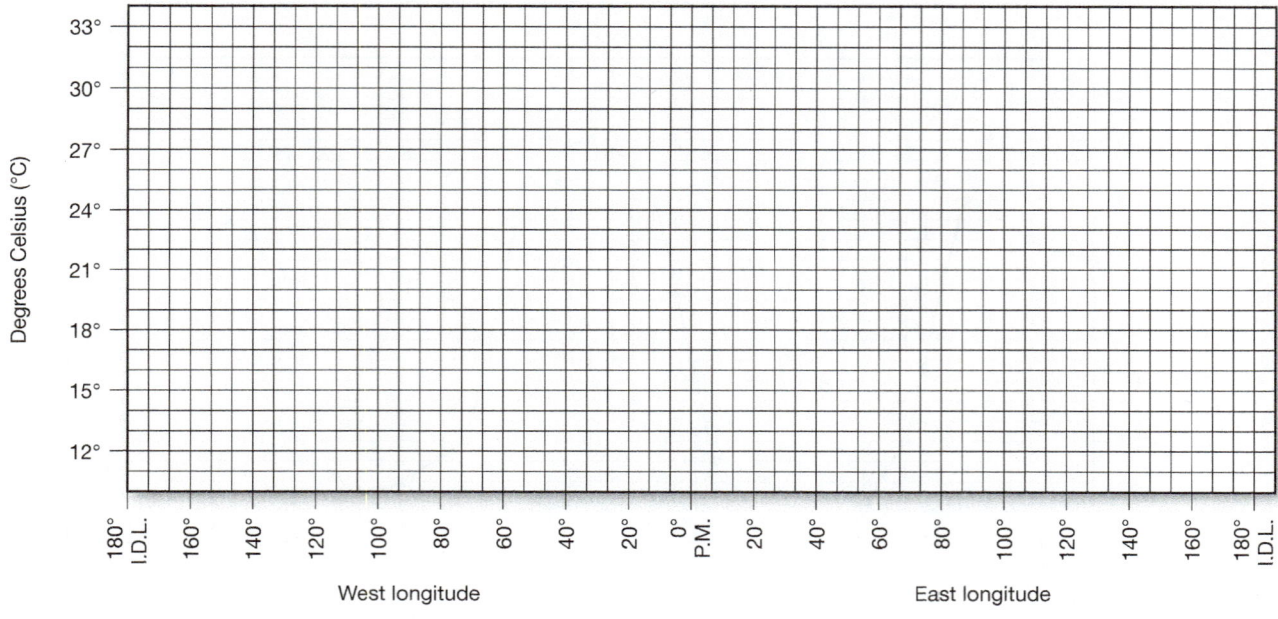

▲ **Figure 9.4a** Plot of average temperatures along 40°N latitude for July

▲ **Figure 9.4b** Plot of average temperatures along 60°E longitude for July

Mastering Geography

Looking for additional lab prep and review materials? Visit the Study Area in Mastering Geography to enhance your geographic literacy and spatial reasoning skills by accessing MapMaster™ interactive maps, videos, *Mobile Field Trips*, *Project Condor* Quadcopter Videos, *In the News* articles, flashcards, web links, self-study quizzes, and other resources. Contact your professor to find out how you can access Mastering Geography.

Video
Exercise 10
Pre-Lab Video

http://goo.gl/HzD3sG

Scan to view the
Pre-Lab video

LAB EXERCISE

10 Earth's Atmosphere: Temperature and Pressure Profiles

The temperature of Earth's atmosphere varies greatly with altitude. The average temperature of Earth's surface is 15°C (59°F), but the temperature of the atmosphere drops to –90°C (–130°F) and climbs to over 1200°C (2200°F). These temperatures are the result of our atmosphere absorbing high-energy radiation from the Sun. The ozone layer converts ultraviolet radiation to long-wave infrared radiation, and the ionosphere absorbs gamma and x-ray radiation. While the temperature at the top of the atmosphere is over 1200°C (2200°F), there is little heat because there are so few molecules at that altitude. The gases that make up air create pressure through their motion, size, and number. This pressure is exerted on all surfaces in contact with the air. The weight of the atmosphere, or **air pressure**, exerts an average force of approximately 1 kg/cm^2 (14.7 lb/in^2) at sea level. Under the acceleration of gravity, air is compressed and therefore denser near Earth's surface. The atmosphere rapidly thins with increased altitude, a decrease that is measurable because air exerts its weight as pressure. Consequently, over half the total mass of the atmosphere is compressed below 5500 m (18,000 ft), 75% is compressed below 10,700 m (35,105 ft), and 90% is compressed below 16,000 m (52,500 ft). All but 0.1% is below an altitude of 50,000 m (163,680 ft), or 50 km (31 mi).

In Lab Exercise 10 we examine Earth's atmospheric temperature and pressure patterns. Lab Exercise 10 features three sections.

Key Terms and Concepts

air pressure
aneroid barometer
barometer
mercury barometer
mesopause

mesosphere
standard atmosphere
stratopause
stratosphere
thermopause

thermosphere
tropopause
troposphere

KEY**LEARNING**concepts

After completion of this lab, you should be able to:

- *Graph* altitudinal temperature changes in the atmosphere.
- *Demonstrate* pressure trends with altitude.
- *Explain* the standard atmosphere concept and *plot* key elements on a graph.
- *Locate* a properly installed barometer (and, if necessary, a thermometer) and *obtain* data over a 3-day period.

Materials/Sources Needed

pencil
colored pencils

calculator
textbook

Lab Exercise and Activities

SECTION 1

Temperature Profile of the Atmosphere

Based on *temperature*, the atmosphere is divided into four distinct zones or layers: the **thermosphere**, **mesosphere**, **stratosphere**, and **troposphere**. The transition area at the top of each temperature region is named using the suffix *-pause*, which means "to cause to change," i.e., **thermopause**, **mesopause**, **stratopause**, and **tropopause**.

Temperatures in the lower atmosphere do not simply decline with altitude, for it is more complex than this and actually varies. In the upper atmosphere the temperature profile shows that temperatures rise sharply in the thermosphere, up to 1200°C (2200°F) and higher. However, we must use different concepts of "temperature" and "heat" to understand this effect. The intense radiation in this portion of the atmosphere excites individual molecules (nitrogen and oxygen) and atoms (oxygen) to high levels of vibration. This *kinetic energy*, the energy of motion, is the vibrational energy stated as "temperature." However, the density of the molecules is so low that little actual heat is produced. Heating in the lower atmosphere near Earth's surface differs because the greater number of molecules in the denser atmosphere transmit their kinetic energy as *sensible heat*, meaning that we can sense it and measure it. **Figure 10.1** gives you the general trend for this temperature profile that you specifically plot in this section using data from **Table 10.1**.

TABLE 10.1 Standard temperature values for the atmosphere		
Boundary of Atmosphere	Altitude	Temperature
Thermopause	480 km (300 mi)	1200°C (2200°F)+
Mesopause	80 km (50 mi)	−90°C (−130°F)
Stratopause	50 km (31 mi)	0°C (32°F)
Tropopause	18 km (11 mi)	−57°C (−70°F)
Surface (No. Hemis.)	Sea level	15°C (59°F)

1. Using the graph in Figure 10.1, plot the standard temperature values given in Table 10.1. (The sea level value has been done for you.) After you plot the data points, connect them with a line graph to complete the profile. Label the layers of the atmosphere and the transition areas at the top of each layer.

   Analysis and completion questions about the standard temperature profile.

2. Briefly explain why the temperature decreases as altitude increases in the troposphere (at the normal lapse rate).

3. Why do temperatures increase throughout most of the stratosphere? Specifically discuss the process that produces this warming effect.

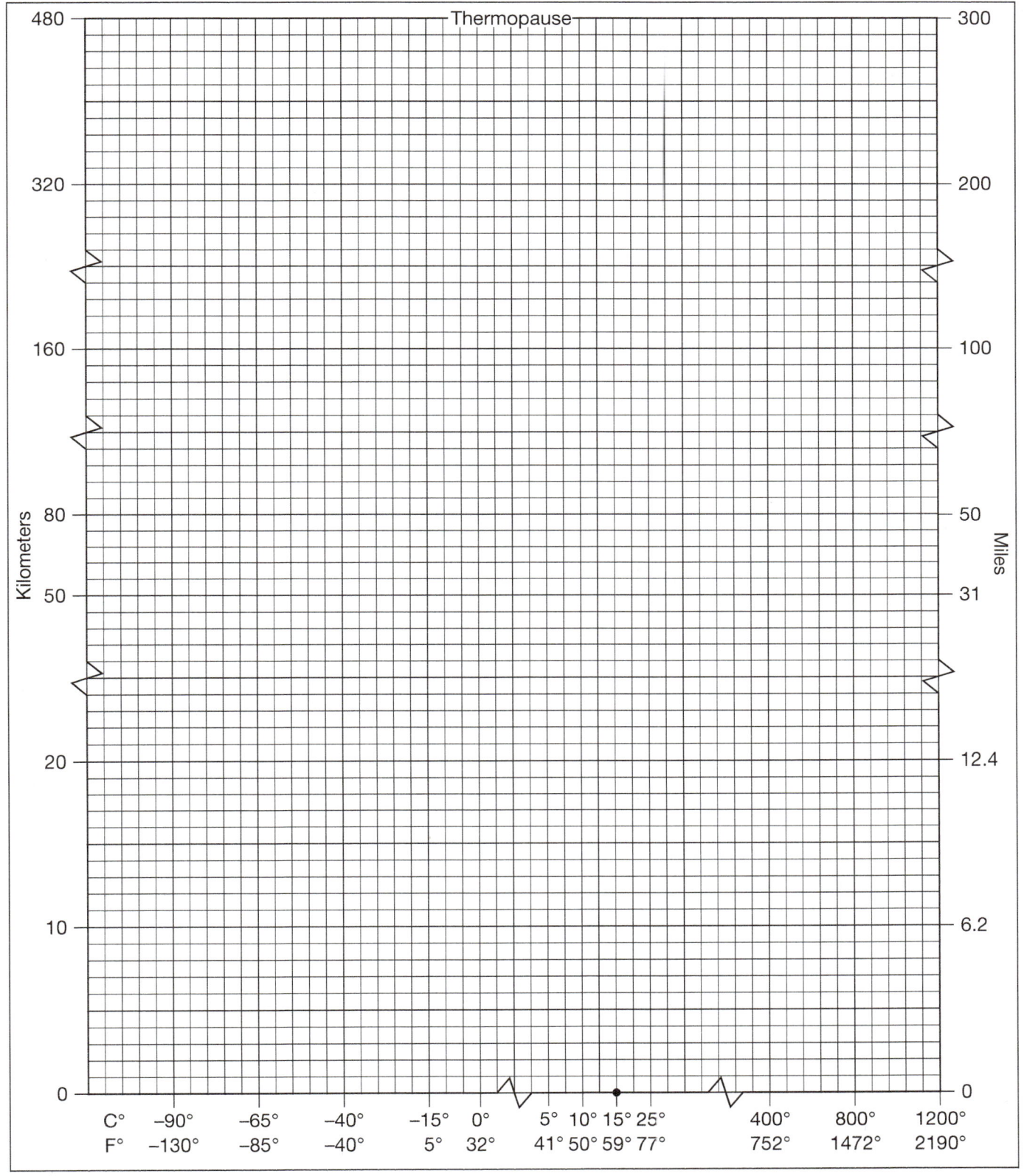

▲ **Figure 10.1** Atmospheric temperature profile graph

4. Briefly explain why temperatures increase in the thermosphere.

SECTION 2

Air Pressure

Any instrument that measures air pressure is called a **barometer**. One type of barometer uses a column of mercury that is counterbalanced by the mass of surrounding air exerting an equivalent pressure on a vessel of mercury to which the column is attached. The **mercury barometer** was developed by Evangelista Torricelli in 1647. Mercury barometers are rarely used in schools because of the toxicity of mercury. A more common type of barometer, the **aneroid barometer**, is a small chamber that is partially emptied of air, sealed, and connected to a mechanism that is sensitive to changes in air pressure. As air pressure varies, the mechanism responds.

Altimetry is the measurement of altitude using air pressure. A pressure altimeter is an instrument that measures altitude based on a strict relationship between air pressure and altitude. Air pressure is measured within the altimeter by an aneroid barometer capsule, with the instrument graduated in increments of altitude. A pilot must constantly adjust the altimeter, or "zero" the altimeter, during flight as atmospheric densities change with air temperature. Another type of altimeter—a radar altimeter—sends and receives radio wavelengths between the plane and the ground to determine altitude.

Normal sea level pressure is expressed as 1013.2 mb (millibars) of mercury (a way of expressing force per square meter of surface area). At sea level standard atmospheric pressure is expressed in several ways:

- 14.7 lb/in^2 (pounds per square inch, or psi)
- 29.9213 inches of mercury (in Hg)
- 1013.250 millibars (mb)
- 101.325 kilopascals (1 kilopascal = 10 millibars)

Some convenient conversions are helpful:

- 1.0 in Hg = 33.87 mb = 25.40 mm Hg = 0.49 lb/in^2
- 1.0 mb = 0.0295 in Hg = 0.75 mm Hg = 0.0145 lb/in^2

The **standard atmosphere** for pressure in millibars and altitude in kilometers is given in **Table 10.2**. This is used in the activity that follows the table.

TABLE 10.2 Standard atmosphere for pressure and altitude			
Altitude (km)	Pressure (mb)	Altitude (km)	Pressure (mb)
0.00	1013.25	10.00	264.99
0.50	954.61	12.00	193.99
1.00	898.76	14.00	141.70
1.50	845.59	16.00	103.52
2.00	795.01	18.00	75.65
2.50	746.91	20.00	55.29
3.00	701.21	25.00	25.49
4.00	616.60	30.00	11.97
5.00	540.48	35.00	5.75
6.00	472.17	40.00	2.87
7.00	411.05	50.00	0.79
8.00	356.51	60.00	0.23
9.00	308.00	70.00	0.06

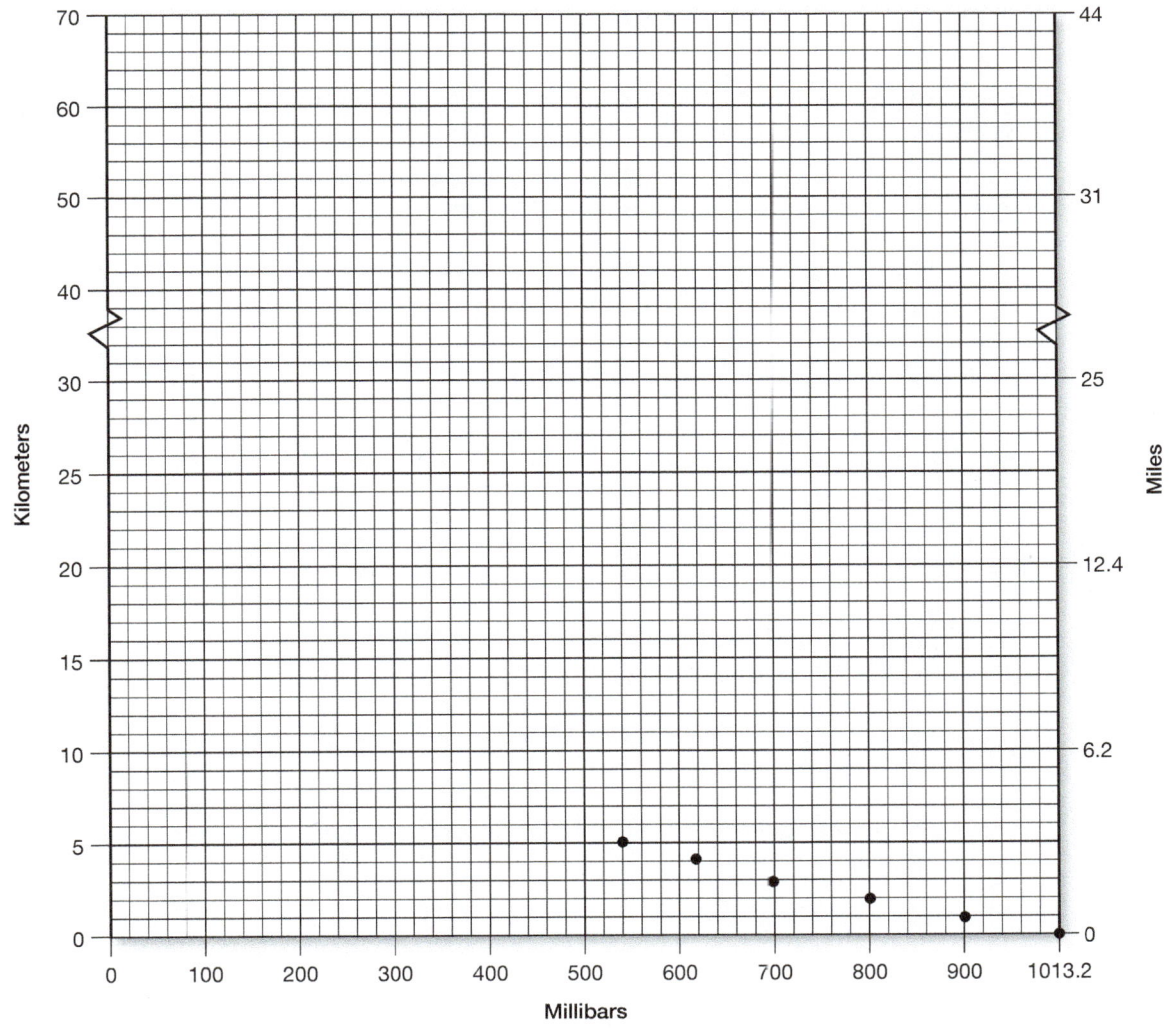

▲ Figure 10.2 Atmospheric pressure profile graph—the standard atmosphere—from the surface to 70 km.

1. Using the graph in **Figure 10.2**, *plot* the standard atmosphere of air pressure decrease with altitude presented in Table 10.2. After completing the plot, connect the data points with a line to complete the pressure profile of the atmosphere. The data points from 0 to 5 km are plotted for you.

2. The information in Table 10.2 allows a determination of the rate of pressure decrease with altitude, which is not at a constant rate. Remember that half of the weight of the total atmosphere occurs below 5500 m (18,000 ft); at that altitude only about half of the total atoms and molecules of atmospheric gases remain to form the mass of the atmosphere. Determine the decrease in pressure between the following altitudes. Express the difference in millibars and inches of mercury. (Conversions are presented earlier in this section.)

1 km interval difference in pressure:

0 and 1 km __[114.49]__ mb; __[3.38]__ in. Hg

2 and 3 km _____ mb; _____ in. Hg

5 and 6 km _____ mb; _____ in. Hg

8 and 9 km _____ mb; _____ in. Hg

9 and 10 km _____ mb; _____ in. Hg

10 km interval difference in pressure:

0 and 10 km __[748.26]__ mb; __[22.07]__ in. Hg

10 and 20 km _____ mb; _____ in. Hg

20 and 30 km _____ mb; _____ in. Hg

40 and 50 km _____ mb; _____ in. Hg

60 and 70 km _____ mb; _____ in. Hg

3. Using the graph you prepared in Figure 10.2, list the *approximate* answers to the following, assuming standard atmosphere conditions. The first answer is provided for you in brackets.

   a) Mount Everest's summit is 8850 m (29,035 ft) above sea level. What is the barometric pressure there, according to the standard atmosphere? [*320 mb*]

   b) Mount McKinley, 6194 m (20,320 ft); air pressure at the summit?

   c) Mount Whitney, 4418 m (14,494 ft); air pressure at the summit?

   d) Yellowstone Lake, Yellowstone N.P., 2356 m (7731 ft); air pressure?

   e) The Petronas Towers I and II, Kuala Lumpur, Malaysia, 452 m (1483 ft); air pressure?

f) In a commercial airliner taking you from San Francisco to New York at 12,000 m (39,400 ft), what percentage of atmospheric pressure is below your plane? What percentage of atmospheric pressure resides above your flight altitude?

4. Why does atmospheric pressure decrease so rapidly with altitude?

SECTION 3

Pressure Readings

Lab Exercise 7, Section 2 asked you to record air temperature and air pressure for 4 days. If you did not record barometric readings at that time, locate a barometer either at the college or university you are attending, or perhaps at home. Many students find that someone in the family was given an aneroid barometer years ago—and perhaps no one quite knows how to use it. If you do not have access to a barometer, find a source for information about barometric pressure (weather broadcasts, Internet sources). At the same time, you may want to record air temperature. As before, the goal is to locate and use a reliable source of barometric pressure. This will be an asset in your life, travel, and activities.

1. You started collecting data on the Weather Calendar in the Prologue Lab. As you learn more about elements of weather, you should begin to apply what you learn to the weather that you experience on a daily basis. Collecting and working with weather data will enhance your awareness and understanding of weather processes.

For at least 4 days, record the air pressure (including the units) and air temperature at approximately the same time of day, if possible. See if you can detect a trend in air pressure changes or a relationship between air pressure and other atmospheric processes. *Record* your observations below.

Date, Pressure, and Temperature	Place of Observation	Time of Day	Barometer Used
Day 1:	Place:	Time:	Type:
Day 2:	Place:	Time:	Type:
Day 3:	Place:	Time:	Type:
Day 4:	Place:	Time:	Type:

2. Weather conditions and state of the sky at the time of your observation:

   Day 1:

   Day 2:

   Day 3:

   Day 4:

3. What have you observed regarding the relationship between air pressure and the observed weather, including the state of the sky?

Mastering Geography

Looking for additional lab prep and review materials? Visit the Study Area in Mastering Geography to enhance your geographic literacy and spatial reasoning skills by accessing MapMaster™ interactive maps, videos, *Mobile Field Trips, Project Condor* Quadcopter Videos, *In the News* articles, flashcards, web links, self-study quizzes, and other resources. Contact your professor to find out how you can access Mastering Geography.

LAB EXERCISE

Earth's Atmosphere: Pressure and Wind Patterns

Differences in air pressure between one location and another produce wind, the horizontal motion of air. The driving forces that produce and determine the speed and direction of surface winds are **gravitational force**, the **pressure gradient force**, the **Coriolis force**, and **friction**. These forces, combined with the energy imbalance from the tropics to the poles, cause the circulation of Earth's atmosphere and ocean currents, including the principal pressure systems and patterns of global surface winds.

The primary high- and low-pressure areas appear on maps of air pressure as cells or uneven belts of similar pressure that stretch across the face of the planet, interrupted by landmasses. The primary winds flow between the primary high- and low-pressure areas. Four broad pressure areas cover the Northern Hemisphere, and a similar set exists in the Southern Hemisphere. Along the equator is a belt of low pressure called the **Intertropical Convergence Zone** or **ITCZ**. Moving away from the equator and toward the poles, there are the **subtropical high-pressure cells**; the **Hawaiian high** in the Pacific Ocean and the Azores or Bermuda high in the Atlantic Ocean. The **subpolar low-pressure cells** are more prominent in winter and include the Aleutian low in the northern Pacific and the Icelandic low in the north Atlantic. **Polar high-pressure cells**, and the winds they drive, are weak because the polar atmosphere receives little energy from the Sun.

Lab Exercise 11 features two sections.

Key Terms and Concepts

Coriolis force
friction
geostrophic winds
gravitational force

Hawaiian high
isobars
ITCZ
polar high-pressure cells

pressure gradient force
subpolar low-pressure cells
subtropical high-pressure cells

KEY**LEARNING**concepts

After completion of this lab, you should be able to:

- *Identify* the four forces that produce wind and *draw* a diagram showing how they create surface and upper atmosphere wind patterns.
- *Interpret* global pressure patterns and *construct* zonal and meridional pressure profiles.
- *Analyze* global pressure patterns by plotting pressure profiles.

Materials/Sources Needed

pencil
colored pencils

calculator
textbook

Lab Exercise and Activities

SECTION 1

Driving Forces Within the Atmosphere and Wind Patterns

The driving forces that produce and determine the speed and direction of surface winds are gravitational force, the pressure-gradient force, the Coriolis force, and friction. Air flows from high pressure to low pressure because of the pressure-gradient force. Without the pressure-gradient force, there would be no wind. The Coriolis and friction forces act only on moving air. The strength of the pressure-gradient force is shown by the spacing of isobars. An **isobar** is an isoline that connects points of equal pressure on a weather map. Closer spacing indicates a steeper pressure gradient and a stronger pressure-gradient force. The Coriolis force makes wind appear to be deflected in relation to Earth's rotating surface. Because Earth rotates eastward, objects appear to curve to the right in the Northern Hemisphere and to the left in the Southern Hemisphere. The Coriolis force is zero along the equator, increases to half the maximum deflection at 30°N and 30°S latitude, and reaches maximum deflection at the poles. Without Coriolis force, winds would move along straight paths between high- and low-pressure areas. Friction force slows wind speeds and reduces Coriolis force close to the surface. Surface friction extends to a height of about 500 m (around 1600 ft). Higher-altitude winds leave the drag of surface friction behind and increase speed. This increases the Coriolis force, producing the pattern of upper-air westerly winds flowing from the subtropics to the poles. Higher-altitude winds do not flow directly from high to low but flow parallel to the isobars around the pressure areas. Such winds are **geostrophic winds** and are characteristic of upper-tropospheric circulation.

At the surface, airflow is initiated by the pressure-gradient force and modified by the Coriolis force and friction, as shown in **Figures 11.1a** and **11.1b**. Above 500 m (1600 ft), airflow is produced by the pressure-gradient force and the Coriolis force, as shown in **Figures 11.1c** and **11.1d**.

Pressure gradient force alone

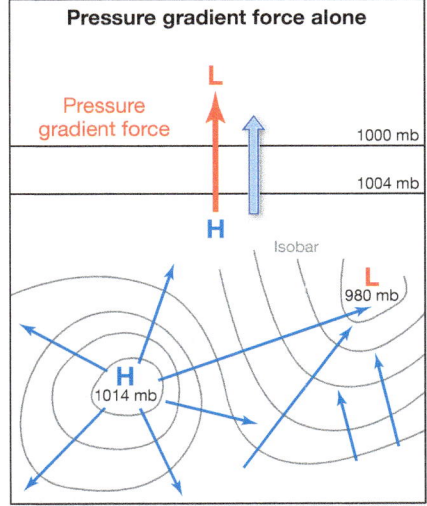

Pressure gradient force

1000 mb

1004 mb

Isobar

L 980 mb

H 1014 mb

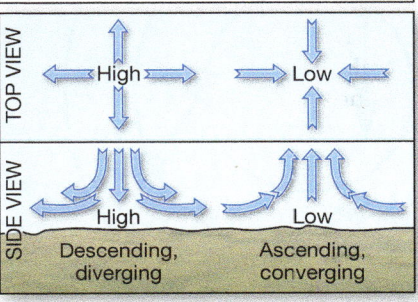

TOP VIEW

High ↔ Low ↔

SIDE VIEW

High — Low

Descending, diverging — Ascending, converging

(a) The pressure gradient force acting by itself on a nonrotating Earth causes winds to move from high to low pressure across isobars.

Pressure gradient + Coriolis forces produce upper-level (geostrophic) winds

Pressure gradient force

1000 mb

Geostrophic wind

1004 mb

Coriolis force

H

L 980 mb

H 1014 mb

Wind

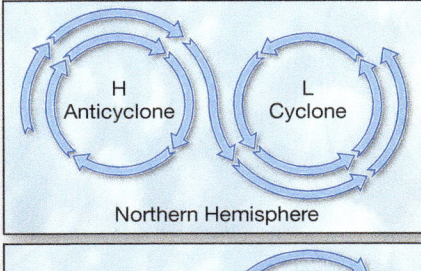

H Anticyclone

L Cyclone

Northern Hemisphere

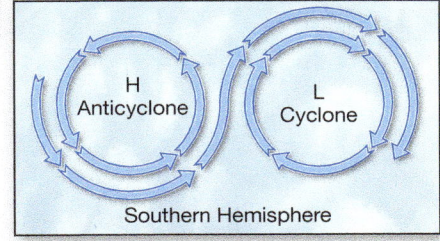

H Anticyclone

L Cyclone

Southern Hemisphere

(b) Earth's rotation adds the Coriolis force, giving a "twist" to air movements so that high-pressure and low-pressure areas develop a rotary motion. Upper-level winds flowing between highs and lows flow parallel to isobars.

Pressure gradient + Coriolis + friction force produce surface winds

Pressure gradient force

L

Surface wind

1000 mb

1004 mb

Frictional force

H

Coriolis force

L 980 mb

H 1014 mb

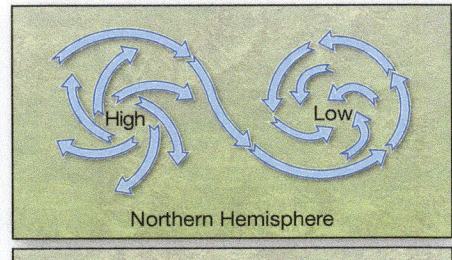

High — Low

Northern Hemisphere

High — Low

Southern Hemisphere

(c) Friction adds a countering force to Coriolis, producing winds that spiral out of a high-pressure area into a low-pressure area. Surface winds cross isobars at an angle. Air flows into low-pressure areas and turns to the left, because of deflection to the right.

▲ **Figure 11.1** Forces that produce winds

Animation
Wind Pattern Development

https://goo.gl/WDXi6c

Animation
Coriolis Force

https://goo.gl/yD0GjD

Animation
Cyclones & Anticyclones

https://goo.gl/WTwTim

1. **Figure 11.2** shows surface wind flow in the Northern and Southern Hemispheres. Place an H or an L in the center of each example to indicate what type of pressure is shown and write the name of the correct hemisphere below each of the examples. Pick one arrow from each example and draw your own arrows to indicate the pressure-gradient force in red, Coriolis force in blue, and friction in brown for that figure, or colors as indicated by your professor, as shown in Figure 11.2a. Label the pressure-gradient force arrow with PGF, the Coriolis force arrow with CF, and the friction with F.

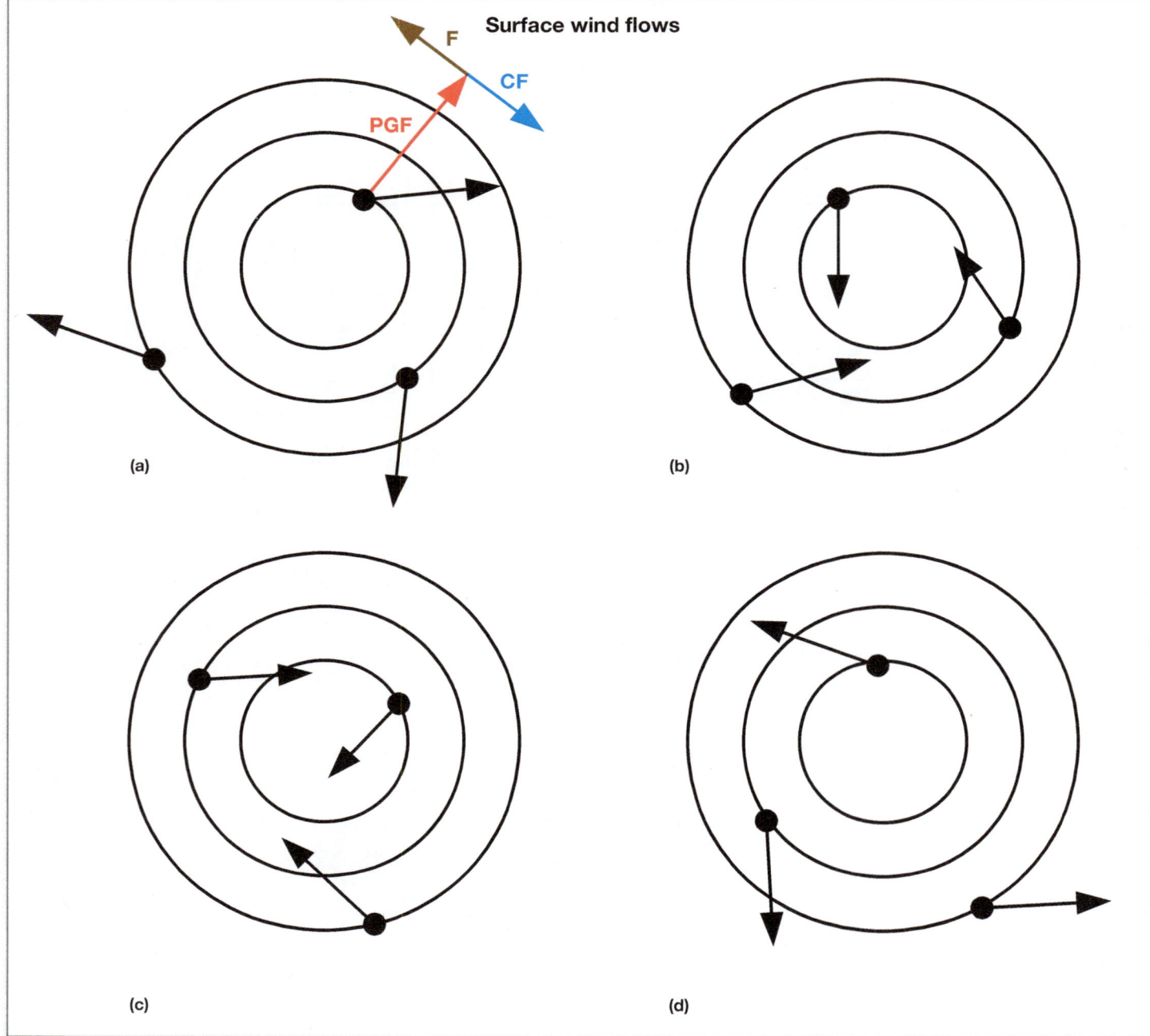

▲ Figure 11.2

of the principal meridian. Note that due to the three sets of imposing mountain ranges, three baselines and meridians are used.

The township and range grid system is based on 6-mile square *townships*, subdivided into 1-mile square *sections*, and quarter-section *homesteads*.

Figure 4.8 portrays this system and shows the derivation of a parcel of land described as "S1/2, NW1/4, NE1/4, SW1/4, Sec. 26, T2S, R4E." Note that you start from the smallest parcel of land ("the south half, of the northwest quarter, of the northeast quarter . . ."). Some land deeds still reflect this system.

Questions and analysis about the township and range system.

1. Complete the fill-ins presented in Figure 4.8 on the next page.

2. Using Figure 4.7 as your guide, which meridian and base line pair would be used to survey the area you are in?

3. As mentioned in the introduction to this section, California uses three different base lines and meridians. List the other states that use more than one base line and meridian pair.

4. If your campus was surveyed using the public-lands system, give a full description of the campus in standard township and range format (consult a topographic map for your campus area).

5. If the state or region where your campus is located was surveyed under a different system (for example, parts of Texas, the original 13 colonies), describe it relative to this system.

Mastering Geography

Looking for additional lab prep and review materials? Visit the Study Area in Mastering Geography to enhance your geographic literacy and spatial reasoning skills by accessing MapMaster™ interactive maps, videos, *Mobile Field Trips, Project Condor* Quadcopter Videos, *In the News* articles, flashcards, web links, self-study quizzes, and other resources. Contact your professor to find out how you can access Mastering Geography.

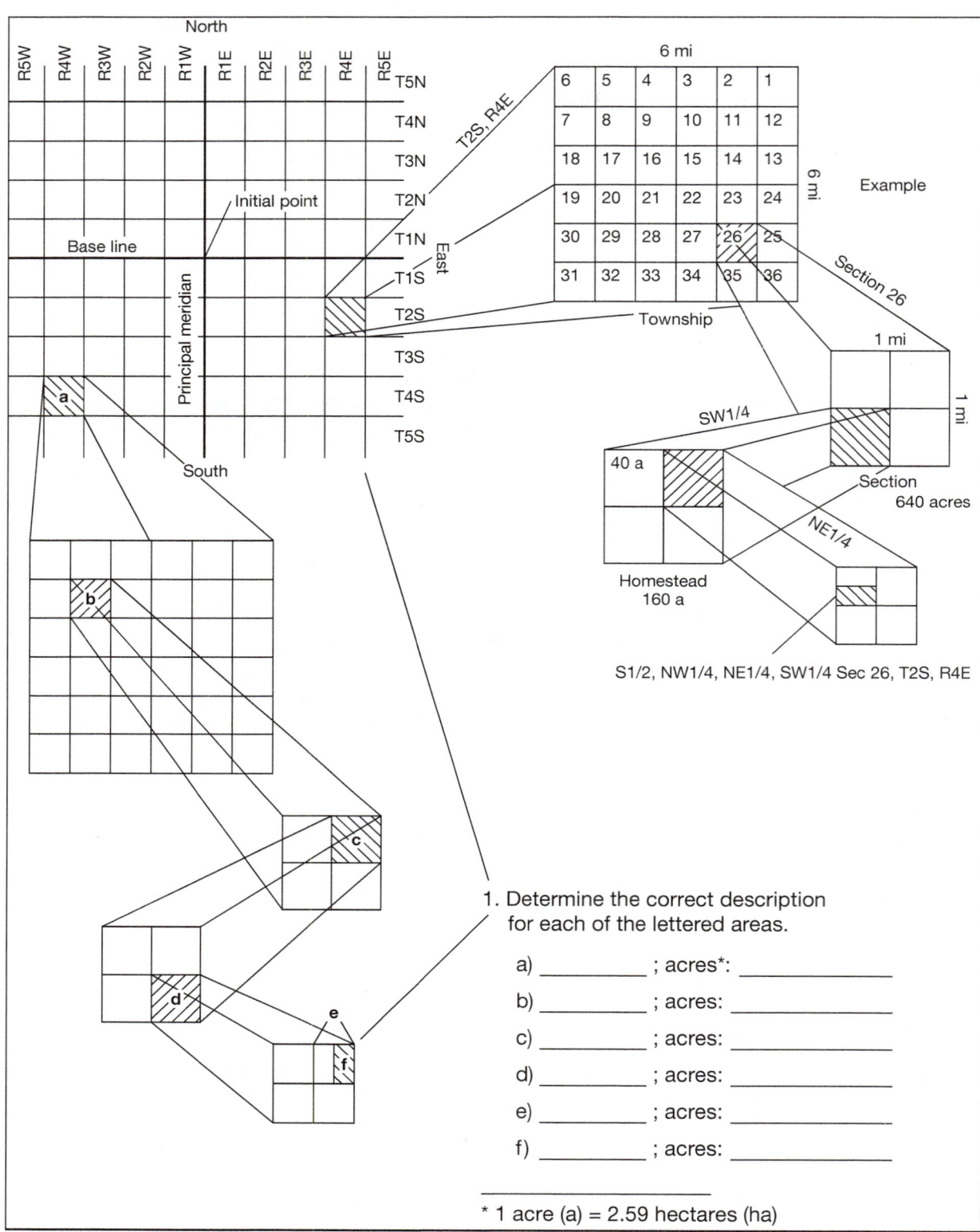

1. Determine the correct description for each of the lettered areas.

   a) _____ ; acres*: _____

   b) _____ ; acres: _____

   c) _____ ; acres: _____

   d) _____ ; acres: _____

   e) _____ ; acres: _____

   f) _____ ; acres: _____

   _____

   * 1 acre (a) = 2.59 hectares (ha)

▲ **Figure 4.8** Township and range rectangular survey grid system illustrating a parcel of land at S1/2, NW1/4, NE1/4, SW1/4, Sec 26, T2S, R4E

Video
Exercise 5
Pre-Lab Video

http://goo.gl/8yj0lS

Scan to view the
Pre-Lab video

LAB EXERCISE

5 Earth–Sun Relationships and Daylength

Earth's distinct seasons are produced by interactions of revolution (annual orbit about the Sun) and rotation (Earth's turning on its axis), Earth's axial tilt (at about 23.5° from a perpendicular to the plane of the ecliptic), axial parallelism (the parallel alignment of the axis throughout the year), and sphericity. As Earth rotates, the boundary that divides daylight and darkness is the **circle of illumination**.

Earth rotates about its axis, an imaginary line extending through the planet from the geographic North Pole to the South Pole. The Tropic of Cancer and Tropic of Capricorn mark the farthest north and south the subsolar point migrates during the year, about 23.5°N and 23.5°S latitude. The **solstices** occur when the Sun is directly over the tropics, and the **equinoxes** occur when the Sun is directly over the equator. The **march of the seasons** refers to the procession of the March equinox, June solstice, September equinox, and the December solstice.

Daylength, the interval between sunrise and sunset, varies during the year, depending on latitude. The equator always receives equal hours of day and night. At 50°N or S latitude, people experience almost 8 hours of annual daylength variation.

At the North and South Poles, the range of daylength is extreme, with a 6-month period of no insolation, beginning with weeks of twilight, then darkness, then weeks of predawn. Following sunrise, daylight lasts for a 6-month period of continuous 24-hour insolation—literally, the poles experience one long day and one long night each year! Lab Exercise 5 has two sections.

Key Terms and Concepts

circle of illumination
daylength

equinox
march of the seasons

solstice
subsolar point

KEY**LEARNING**concepts

After completion of this lab, you should be able to:

- *Label* diagrams representing Earth–Sun relations in the annual march of the seasons.
- *Calculate* daylength for selected locations for the solstices and equinoxes and *analyze* the relationship between latitude and seasonality.

Materials/Sources Needed

pencil
calculator

color pencils
ruler

Lab Exercise and Activities

SECTION 1

Earth–Sun Relations—Seasonality

Because Earth is spherical, its curved surface presents a continually varying angle to the incoming parallel rays of insolation. Differences in the angle of solar rays at each latitude result in an uneven distribution of insolation and heating. The place receiving maximum insolation is the point where insolation rays are perpendicular to the surface (radiating from directly overhead), called the **subsolar point**. All other places receive insolation at less than a 90° angle and thus experience more diffuse energy receipts.

Solar beam angles become more pronounced at higher latitudes. As a result, during a year's time, the top of the atmosphere above the equatorial region receives 2.5 times more insolation than that received above the poles.

Animation
Earth Sun Relations

http://goo.gl/LLDpgQ

1. On the accompanying diagrams in **Figure 5.1**, complete the following items. June has been started for you. Also, you may want to consult your physical geography text for reference.

   a) Extend the Sun's rays until they *intersect* (pass through) or are *tangent* to Earth's surface (touch at only one point). (See the June solstice Earth profile diagram for examples.)

   b) Where the rays intersect or are tangent to Earth's surface, draw a short line tangent to the surface, indicating the angle at which the Sun's rays are intercepted.

   c) Label the vertical rays (striking Earth at a 90° angle) and tangent rays (parallel to Earth) with their latitude for the December solstice and the March equinox, as shown for the June solstice. Label the subsolar point for June, March, and September.

   d) Draw the following on the Earth profiles and label:

North and South Poles	Circle of illumination
Equator	Arctic and Antarctic Circles
Tropics of Cancer and Capricorn	

   e) Lightly shade the portion of Earth that is experiencing night, or the night half of the circle of illumination. Half of Earth is in sunlight and half is in darkness at any moment. The traveling boundary that divides daylight and darkness is called the circle of illumination—it is the day–night dividing circle.

2. On the equinoxes, the subsolar point is at the equator, and the circle of illumination runs through

   _____ °N and _____ °S. Therefore, the far north and south latitudes of the circle of illumination are _____° away from the subsolar point. On the June solstice, the subsolar point is at the Tropic of Cancer at _____° _____, and the circle of illumination passes through _____ °N and _____ °S. Which latitudes does the circle of illumination pass through on the June solstice? On February 26 the subsolar point is at 9°S, and the circle of illumination passes through _____ °N and _____ °S.

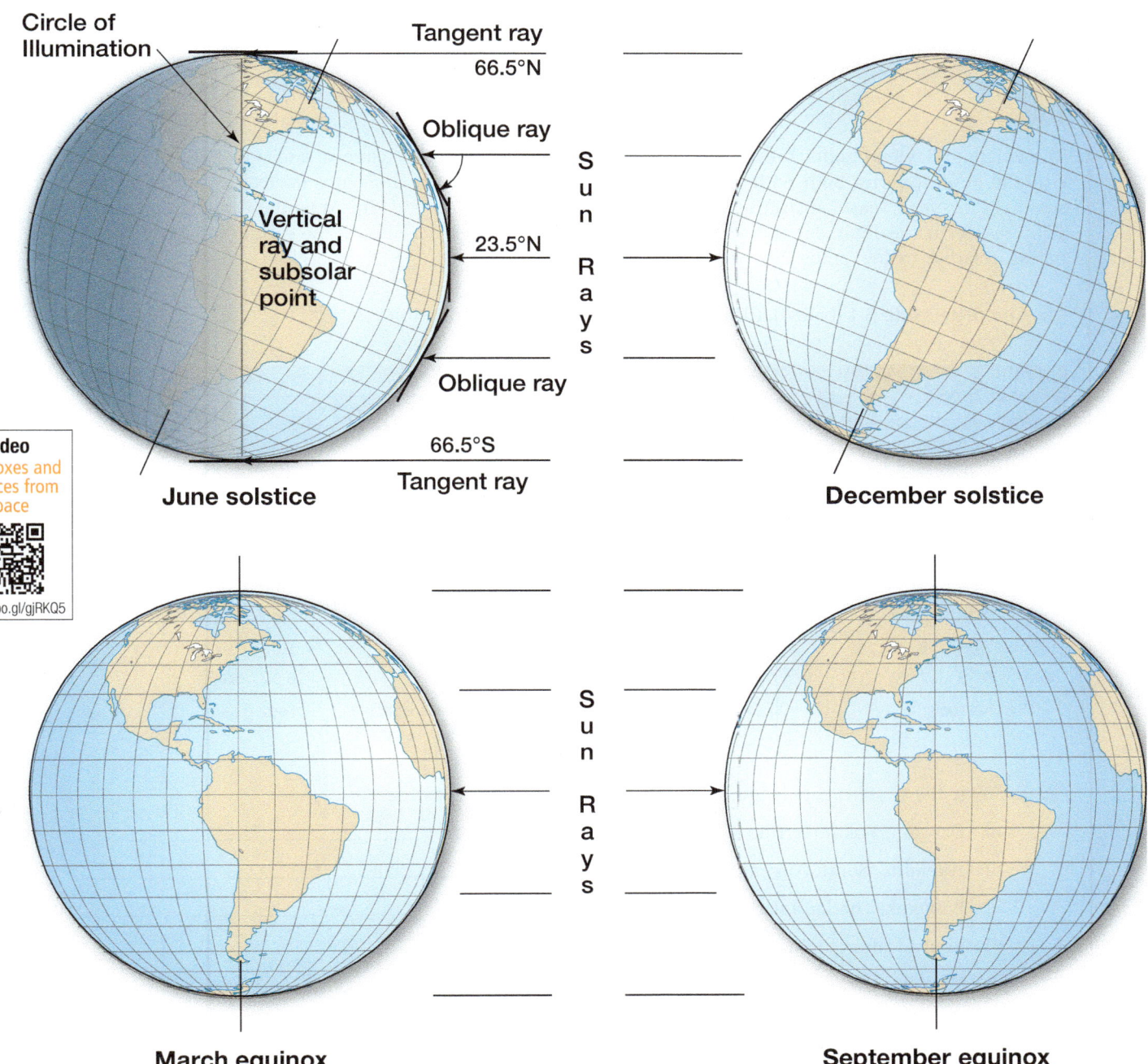

▲ **Figure 5.1** Earth–Sun relationships

SECTION 2

Daylength

The length of daylight hours, the time between sunrise and sunset, varies both with latitude and the season of the year. Seasonal shifting of the subsolar point results in a realignment of the circle of illumination relative to the poles. See the diagrams in Section 1 of this exercise and note the proportions of each latitude line (equator, Tropics of Cancer and Capricorn, Arctic and Antarctic Circles) that are in daylight vs. darkness at the various seasons.

1. Complete **Table 5.1** (below), filling in the daylength at selected latitudes.

TABLE 5.1 Daylength—the time between sunrise and sunset—at selected latitudes for the Northern Hemisphere

	Winter Solstice (December Solstice) December 21–22			Vernal Equinox (March Equinox) March 20–21			Summer Solstice (June Solstice) June 20–21			Autumnal Equinox (September Equinox) September 22–23		
	A.M.	P.M.	Daylength	A.M.	P.M.	Daylength	A.M.	P.M.	Daylength	A.M.	P.M.	Daylength
0°	6:00	6:00	_____	6:00	6:00	_____	6:00	6:00	_____	6:00	6:00	_____
30°	6:58	5:02	_____	6:00	6:00	_____	5:02	6:58	_____	6:00	6:00	_____
40°	7:26	4:34	_____	6:00	6:00	_____	4:34	7:26	_____	6:00	6:00	_____
50°	8:05	3:55	_____	6:00	6:00	_____	3:55	8:05	_____	6:00	6:00	_____
60°	9:15	2:45	_____	6:00	6:00	_____	2:45	9:15	_____	6:00	6:00	_____
90°	No sunlight			Rising Sun			Continuous sunlight			Setting Sun		

2. Estimate the approximate length of daylight for the following locations:

    a) Dawson, Yukon Territory, Canada (64° N), on December 21 _____

    b) Adelaide, South Australia (35° S), on June 21 _____

    c) Bangkok, Thailand (14° N), on March 20 _____

    d) Your location on June 21 _____

    e) Your location on December 21 _____

3. Complete the following graph, using the values for daylength calculated in Question 1 for the following latitudes: 0°, 30°, 40°, 50°, 60°, and 90°. Use a different color for each latitude. The line for 30° has already been done for you.

▲ **Figure 5.2** Daylength and latitude

4. Explain how changes in the Sun's angle above the horizon and daylength vary with the seasons at the equator, 30°, 60°, and 90°. What is the general relationship between latitude and the amount of seasonal change?

Mastering Geography

Looking for additional lab prep and review materials? Visit the Study Area in Mastering Geography to enhance your geographic literacy and spatial reasoning skills by accessing MapMaster™ interactive maps, videos, *Mobile Field Trips*, *Project Condor* Quadcopter Videos, *In the News* articles, flashcards, web links, self-study quizzes, and other resources. Contact your professor to find out how you can access Mastering Geography.

NOTES

Video
Exercise 6
Pre-Lab Video

http://goo.gl/mcZcpV
Scan to view the
Pre-Lab video

LAB EXERCISE

Insolation and Seasons

6

Earth's systems are powered by a constant flow of **insolation**—incoming solar radiation. Earth intercepts only one two-billionth of the entire solar output, yet this is the singular significant source of energy for living systems. This exercise examines the nature of this energy and contrasts the solar output with energy re-radiated by Earth's surface back to space. The Sun's rays arrive at the top of the atmosphere in parallel beams, but the curvature of Earth presents surfaces at differing angles to this incoming radiation: Lower latitudes receive more direct (vertical) illumination, whereas higher latitudes receive more indirect (oblique) rays. This pattern of uneven heating produces an energy imbalance in the atmosphere and on Earth's surface below—namely, equatorial and tropical energy surpluses and polar energy deficits.

Earth's systems are further influenced by shifting seasonal rhythms produced by changing day-length and Sun altitude (height of the noon Sun above the horizon at a given location). These seasonal changes become more noticeable toward the poles. Seasons occur as a result of Earth–Sun relationships produced by **rotation**, **revolution**, tilt of the axis, axial parallelism, and Earth's sphericity. Lab Exercise 6 has two sections.

Key Terms and Concepts

altitude
analemma
declination

insolation
revolution

rotation
zenith

KEY**LEARNING**concepts

After completion of this lab, you should be able to:

- *Identify* the pattern of daily energy receipts at the top of the atmosphere at different latitudes throughout the year and *graph* the observations.

- *Determine* the angle of incidence of insolation and the resultant intensity of the solar beam for various latitudes.

- *Utilize* the analemma to *determine* the subsolar point, latitude, and noon Sun angle (altitude) for several locations using the analemma.

Materials/Sources Needed

pencil
calculator

color pencils
ruler

Distribution of Insolation at the Top of the Atmosphere

As discussed in Lab Exercise 5, Earth's curved surface produces an uneven distribution of energy across the latitudes. Only the subsolar point receives insolation from directly overhead: All other locations receive the Sun's rays at increasingly lower angles as their distance increases from the subsolar point. Lower sun angles produce less intense energy at the surface. The Sun's rays striking a surface at a 30° angle are spread out over twice the surface area as that receiving a perpendicular beam.

The greater length of travel through the atmosphere at higher latitudes reduces solar intensity due to increased scattering within the atmosphere. This energy receipt also varies daily, seasonally, and annually as Sun angle and daylength vary less toward the equator, more toward the poles. The graph in **Figure 6.1** plots the daily variation in insolation for selected latitudes at the top of the atmosphere (480 km, 300 mi). Latitudes are marked along the left side in 10° intervals. The months of the year are marked and labeled across the top and bottom. The dashed curved line shows the declination of the Sun (the latitude of the

▲ **Figure 6.1** Total daily insolation received at the top of the atmosphere charted in watts per square meter by latitude and month (1 watt/m² = 2.064 cal/cm²/day). Adapted from the Smithsonian Institution Press from *Smithsonian Miscellaneous Collections: Smithsonian Meteorological Tables*, vol. 114, 6th Edition. Robert List, ed. Smithsonian Institution, Washington, DC, 1984, p. 419, Table 134.

subsolar point) throughout the year. To read the graph, select a latitude line and follow it from left to right through the months of the year. The daily insolation totals are read from the curved lines, given in watts per square meter per day (W/m^2 per day). For example, at the top of the atmosphere above 30°N latitude the insolation values (in W/m^2 per day) are as follows (readings are approximate extrapolate when necessary): 240 on January 1, 250 on January 15, and 275 on February 1.

1. Using the data in Figure 6.1 and the graph provided below, plot data for specific latitudes for the first of each month. The graph you create will enable you to note the changes throughout the year in the amount of insolation received at a given latitude. Compare the differences in annual insolation patterns at various latitudes. Plot all seven latitudes on one graph, using color pencils to distinguish each. (Be sure to include a legend.)

   - North Pole (started on the graph with a red line and xs)
   - New York (started on the graph; with a blue line and dots)
   - Tropic of Cancer
   - Equator
   - South Pole

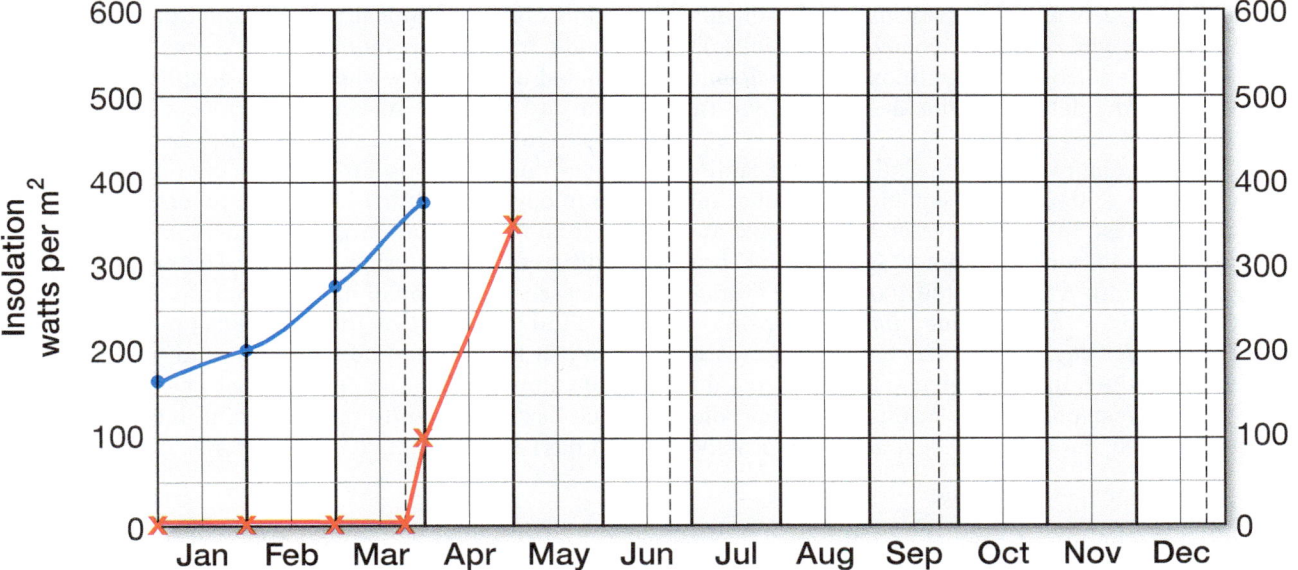

After completing the graph of the energy receipt at these different latitudes, complete the following:

2. Compare and contrast the plotted insolation values at the equator and either of the poles. What factors explain these different patterns of energy receipt?

3. What factor results in the June (summer) solstice energy receipt at the North Pole exceeding that received on the same day at the equator?

4. What is happening at the poles on the March (vernal) equinox?

5. What kind of generalization can you make about the relationship between latitude and annual variation in insolation?

6. **Challenge question:** Why do you think the South Pole receives over 550 W/m^2 at the December solstice, whereas the North Pole receives over 500 W/m^2 during the June solstice?

SECTION 2

Seasonal Variation in the Sun's Declination and Altitude

Seasonal variations are a response to changes in the Sun's **altitude**, the angle between the horizon and the noon Sun. The Sun's **declination**, the latitude of the subsolar point, migrates annually through 47° of latitude—between the *Tropic of Cancer* at 23.5°N and the *Tropic of Capricorn* at 23.5°S. The **analemma** is a convenient device to track the passage of the Sun's path and declination throughout the year (**Figure 6.2**). The horizontal lines are latitudes from 25°N to 25°S. The "figure 8" is a calendar, with each day of the year represented by a black or white segment.

Find August 20 on the analemma; note that it lies on the 12°N line—the subsolar point on that date. While you should remember the subsolar points for the solstices and equinoxes, an analemma will help you determine the declination for the other days of the year.

The analemma also is useful for ascertaining the positive and negative equations of time and periods of "fast-Sun times" (greatest in October and November) and "slow-Sun times" (greatest in February and March). A 24-hour (86,400-second) average day determines *mean solar time*. However, an *apparent solar day* is based on observed successive passages of the Sun over a given meridian. Any difference between observed solar time and mean solar time is called the *equation of time*. On successive days, if the Sun arrives overhead at a meridian *after* 12:00 noon local standard time (taking longer than the 24 hours of a mean solar day)—like an airliner arriving later than scheduled—the equation of time is *negative*, and the Sun is described as "slow." If, on the other hand, the Sun arrives overhead *before* 12:00 noon local time on successive days (taking less time than 24 hours)—like an airliner arriving ahead of schedule—the equation of time is *positive*, and the Sun then is described as "fast." The combination of the tilt and the eccentricity of Earth's orbit around the sun produces this difference between mean solar time and apparent solar time. Because Earth's orbit is elliptical rather than circular, Earth actually travels more quickly when we are at perihelion and more slowly when we are at aphelion. The solar day increases in length by up to 7.9 seconds for several months and then decreases in length for several months. This difference accumulates from day to day and results in the equation of time ranging from up to 16 minutes slow and 14 minutes fast at different times during the year. The equation of time is across the top of the analemma.

Find January 17 on the analemma (marked with a dot). It lies on the line labeled 9 minutes, on the "Sun slow" side. This means that the Sun will reach its "noon" **zenith** (highest point in the sky on that day) at 12:09 P.M., 9 minutes "late."

1. Use the analemma to find:

    a) The subsolar point on November 10 (marked) _____ *[17° south latitude]* _____

    b) The subsolar point on May 11 _____

    c) The subsolar point today _____

    d) The date(s) when the declination is 9°S _____

    e) The date(s) when the declination is 21°N _____

2. At what clock time does the Sun actually reach zenith on:

    October 13? ____ *[11:47 A.M. The Sun is 13 minutes fast, so it will reach its zenith at 11:47 A.M.]* ____

    March 8? _____

    May 20? _____

    Today? _____

As we have noted, the noon Sun angle changes at any given latitude throughout the year. Using the analemma to determine the subsolar point (Sun's declination), you can calculate the altitude of the noon Sun (Sun ∠) for any location by using the following formula:

$$\angle = 90° - (\text{arc distance between your latitude and subsolar point})$$

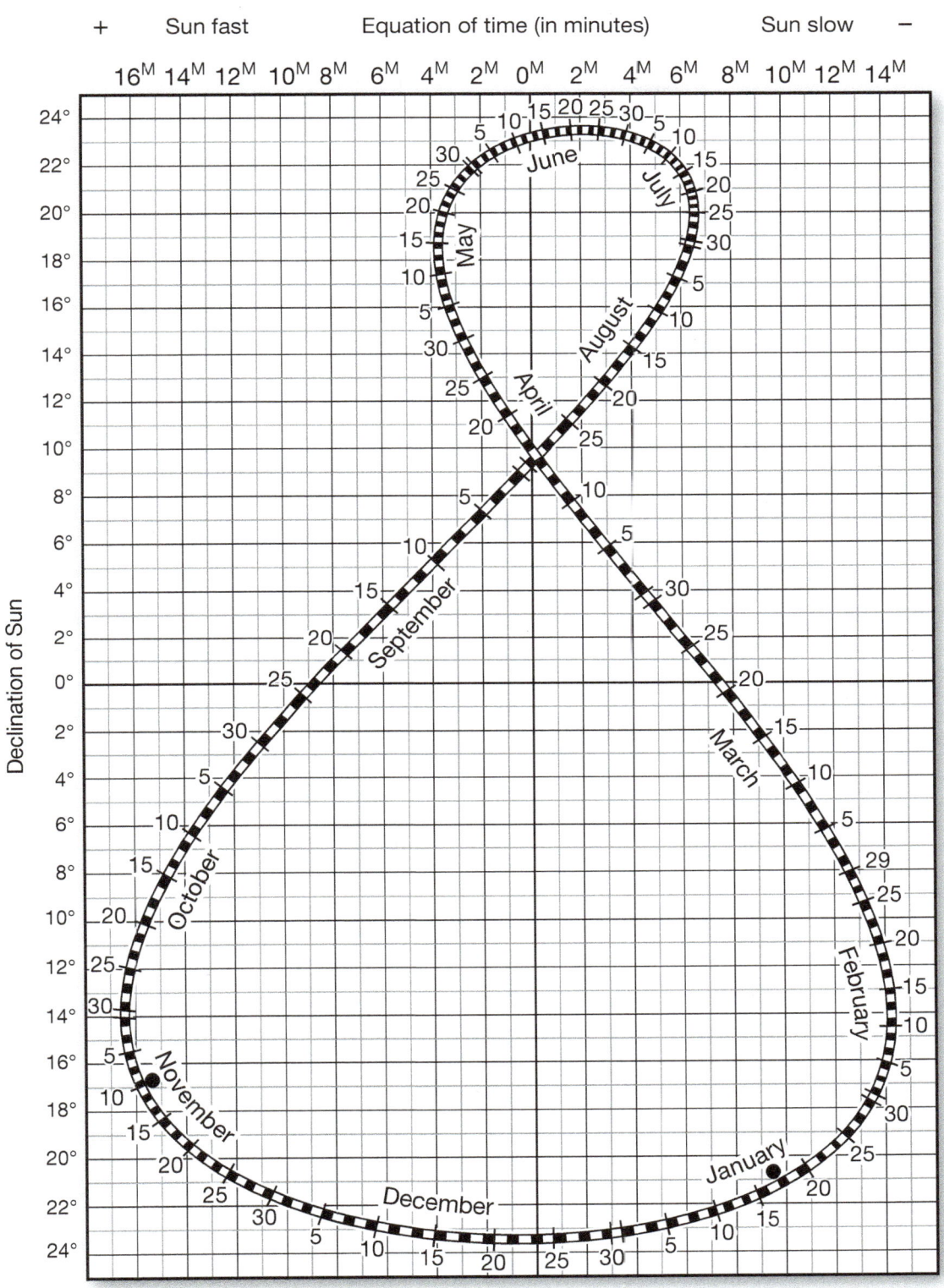

▲ **Figure 6.2** The analemma

EXAMPLES:

You are at 30°N on December 21. You know that on the December solstice, the subsolar point of the noon Sun is 23.5°S. Using the above formula:

$\angle = 90° -$ (arc distance between your latitude and subsolar point)

$\angle = 90° - (30°N \leftrightarrow 23.5°S)$(*the symbol $\leftrightarrow$ is to indicate $\leftrightarrow$ "arc distance between"*)

$\angle = 90° - 53.5°$

$\angle = 36.5°$ above the southern horizon from your location

You determined that on December 21, if you are standing at 30° north latitude, you observe the Sun's noon altitude at 36.5° above the southern horizon (since you are north of the subsolar point on this day).

If you are at 40°N on April 20, you determine from the analemma that the subsolar point is 10°N; therefore:

$\angle = 90° -$ (arc distance between your latitude and subsolar point)

$\angle = 90° - (40°N \leftrightarrow 10°N)$

$\angle = 90° - 30°$

$\angle = 60°$ above the southern horizon*

* Note: You would observe the same Sun altitude on August 25, when the subsolar point is once again at 10°N. If you end up with a negative value for the Sun altitude, then the Sun is below the horizon and will not be visible on that day—as someone might experience north of the Arctic Circle in the Northern Hemisphere winter.

3. You are vacationing at Disney World near Orlando, Florida (28.5°N), on June 3. At what altitude will you observe the noon Sun? Include correct horizon. (Show work.) The problem has been started for you.

   $\angle = 90° -$ (arc distance between your latitude and subsolar point)

   $\angle = 90° - (28.5°N \leftrightarrow$ subsolar point)

4. On that same day, friends of yours are sightseeing at Iguaçu Falls in Brazil (25.7°S). At what altitude will they observe the noon Sun?

5. Calculate the altitude of the noon Sun if you are vacationing in Kuala Lumpur, Malaysia (3.2°N), on July 25.

6. Suppose you were going to install photovoltaic solar panels to generate electricity from the Sun at your house, and you wanted the Sun to shine on them at a 90° angle on the equinoxes. At what angle would they need to be placed, based on the latitude of your home? Should they face north or south?

Whether you view the noon Sun above your northern or your southern horizon depends upon your latitude. The subsolar point is limited to the latitudinal belt between the Tropics of Cancer and Capricorn. An observer who is poleward of the tropics will see the noon Sun only above the "equatorward" horizon (above the southern horizon if north of the Tropic of Cancer and above the northern horizon if south of the Tropic of Capricorn). An observer on or between the tropics will measure the noon Sun above the northern horizon part of the year, above the southern horizon part of the year, and directly overhead on 1 or 2 days.

Use the analemma in Figure 6.2 to answer the following questions:

7. **a)** If you are at 5°N latitude, on what dates will the noon Sun be directly overhead?

  b) Approximately how many days will you view the noon Sun above your northern horizon? _____

      … above your southern horizon? _____

  c) If you are at 18°S latitude, on what dates will the noon Sun be directly overhead?

  d) Approximately how many days will you view the noon Sun above your northern horizon? _____

      … above your southern horizon? _____

Using the same relationship between latitude and Sun altitude, you can also determine your latitudinal location by measuring the solar noon altitude, which can be done with a sextant or more simply with a protractor and a stick. For example:

    Using a sextant, you measure the noon Sun angle at 59° above the southern horizon on March 11. From the analemma, you know that the subsolar point is 4°S; therefore:

$$\angle = 90° - \text{(arc distance between your latitude and subsolar point)}$$

$$59° = 90° - \text{(your latitude} \leftrightarrow 4°\text{S)}$$

$$59° = 90° - 31°$$

If the arc distance between your latitude and the subsolar point = 31°, and the subsolar point is 4°S, then your latitude must be 27°N (31° − 4°). (You must be north latitude because the Sun is viewed above your southern horizon.)

    This method of determining latitude, used along with chronometers and local time to determine longitude (Lab Exercise 1), allowed sailors or other navigators to obtain their absolute global location in the days before GPS (Global Positioning System) units were devised.

8. Calculate the latitude of a mariner who has measured the noon Sun altitude as 60° above the southern horizon on August 13.

9. What would the mariner's latitude be if the Sun had been measured above the northern horizon?

10. Calculate the noon Sun angle for the following locations for the solstices and the equinoxes (March 20 or September 22, June 20, and December 21). Then, use the following graphs with your protractor to draw a side view of the noon Sun angle for each of the two solstices and one of the equinoxes. The graph for La Ronge, Saskatchewan is drawn for you and is shown with a protractor. If needed, refer to the analemma to determine the Sun's declination on these dates for each of the four locations.

a) Arctic Circle (66.5°N)

June solstice: December solstice: Equinoxes:

b) La Ronge, Saskatchewan (55°N)—completed for you

June solstice: 58.5°   December solstice: 11.5°   Equinoxes: 35.0°

c) Atlanta, Georgia (34°N)

June solstice: December solstice: Equinoxes:

d) Equator (0.0°)

June solstice: December solstice: Equinoxes:

e) Tropic of Capricorn (23.5°S)

June solstice: December solstice: Equinoxes:

f) Dunedin, New Zealand (46°S).

June solstice: December solstice: Equinoxes:

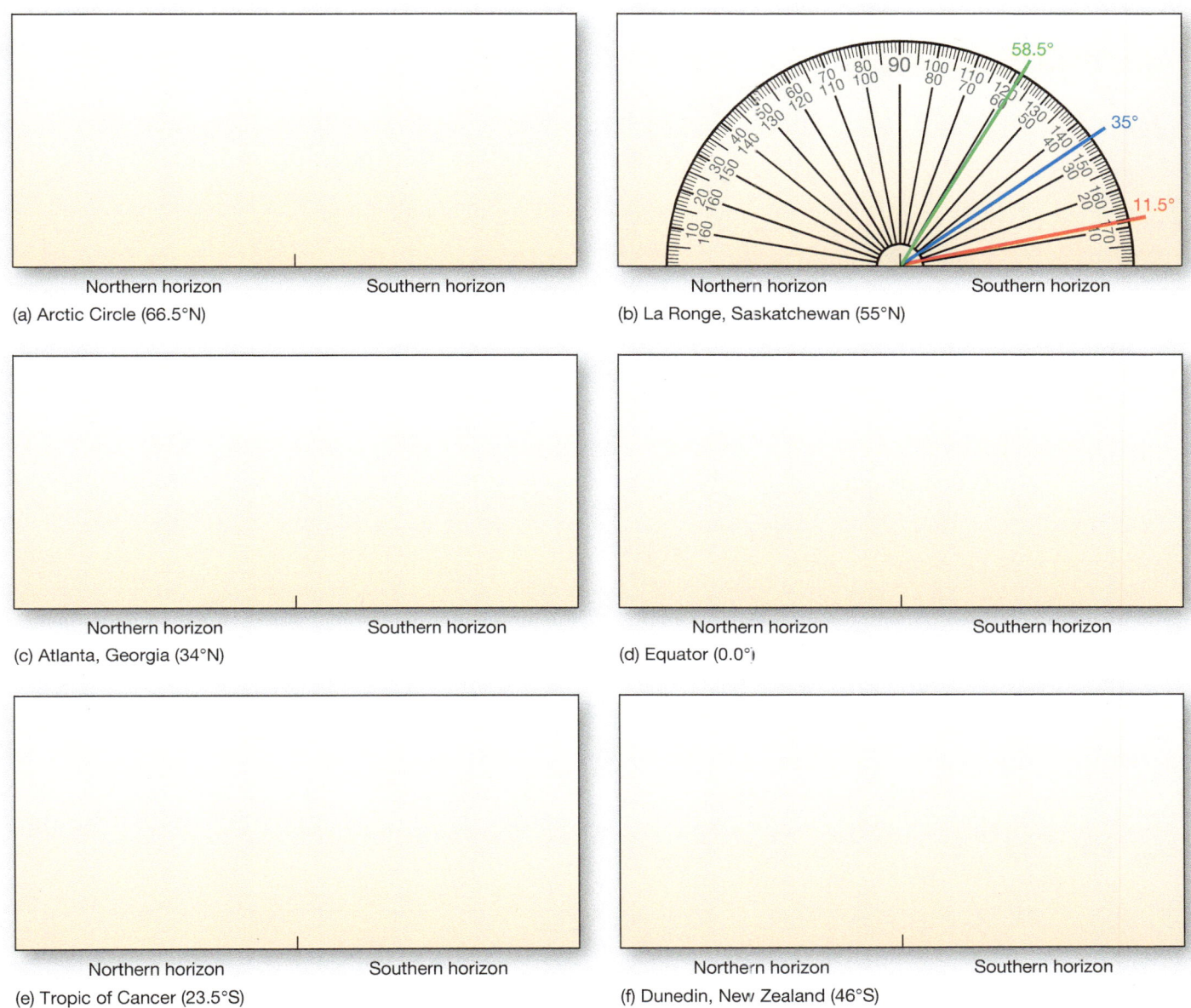

Northern horizon   Southern horizon

(a) Arctic Circle (66.5°N)

Northern horizon   Southern horizon

(b) La Ronge, Saskatchewan (55°N)

Northern horizon   Southern horizon

(c) Atlanta, Georgia (34°N)

Northern horizon   Southern horizon

(d) Equator (0.0°)

Northern horizon   Southern horizon

(e) Tropic of Cancer (23.5°S)

Northern horizon   Southern horizon

(f) Dunedin, New Zealand (46°S)

Mastering Geography

Looking for additional lab prep and review materials? Visit the Study Area in Mastering Geography to enhance your geographic literacy and spatial reasoning skills by accessing MapMaster™ interactive maps, videos, *Mobile Field Trips*, *Project Condor* Quadcopter Videos, *In the News* articles, flashcards, web links, self-study quizzes, and other resources. Contact your professor to find out how you can access Mastering Geography.

NOTES

Name: _____   **Laboratory Section:** _____

Date: _____   **Score/Grade:** _____

Video
Exercise 7
Pre-Lab Video

http://goo.gl/OSSrkb

Scan to view the
Pre-Lab video

LAB EXERCISE

Temperature Concepts

Temperature, a measure of sensible heat energy present in the atmosphere and other media, indicates the average **kinetic energy**, the energy of motion of individual molecules within the atmosphere. You observed air temperature as part of the "Weather Calendar" exercise in the Prologue Lab.

A variety of temperature regimes worldwide affect cultures, decision making, and resource consumption. Global temperature patterns are changing in a warming trend and are the subject of much scientific, geographic, and political interest. In Lab Exercise 7, we examine some temperature concepts and convert between Celsius and Fahrenheit. The effect of temperature in combination with wind and humidity is also presented, since wind chill and the heat index produce apparent temperatures that represent a risk in regions of North America. Lab Exercise 7 features three sections.

Key Terms and Concepts

apparent temperature (sensible temperature)   kinetic energy   temperature
heat index   sensible heat   wind chill factor

KEY**LEARNING**concepts

After completion of this lab, you should be able to:

- **Differentiate** between and convert metric and English units of temperature measures.
- **Obtain** temperature data over a 3-day period.
- **Determine** wind-chill factors for various temperature and wind situations and heat index factors for various temperature and humidity situations.

Materials/Sources Needed

pencil   highlighter pen
color pencils   calculator

Lab Exercise and Activities

SECTION 1

Temperature Concepts, Terms, and Measurements

The amount of heat energy present in any substance is measured and expressed as its temperature. Temperature actually is a reference to kinetic energy, the speed at which atoms and molecules that make up a substance are moving. Heat that we measure with a thermometer is **sensible heat**. Changes in temperature are caused by the absorption (gain) or emission (loss) of energy. The temperature at which all motion in a substance stops is called 0° absolute temperature. Its equivalent in different temperature-measuring schemes is −273° Celsius (C), −459.4° Fahrenheit (F), or 0 Kelvin (K).

The Celsius scale (formerly called centigrade) is named after Swedish astronomer Anders Celsius (1701–1744). It divides the difference between the freezing and boiling temperatures of water into 100 degrees on a decimal scale, with freezing at 0° and boiling at 100°.

The Fahrenheit scale places the freezing point of water at 32°F (0°C, 273°K) and boiling point of water at 212°F (100°C, 373°K), with 180 F° difference between freezing and boiling. The United States remains the only major country still using the Fahrenheit scale, named after the German physicist Daniel Fahrenheit (1686–1736), who invented the alcohol and mercury thermometers and who developed this thermo-metric scale in the early 1700s. Most countries use the Celsius scale to express temperature.

The Kelvin scale, named after British physicist Lord Kelvin (born William Thomson, 1824–1907) who proposed the scale in 1848, is used in science because temperature readings are proportional to the actual kinetic energy in a material.

The formulas to use to convert between Celsius and Fahrenheit temperatures are:

To convert °C to °F:

$$°F = (°C × 1.8) + 32$$
$$\text{or} \quad °F = 1.8°C + 32$$

To convert °F to °C:

$$°C = (°F − 32) ÷ 1.8$$
$$\text{or} \quad °C = (°F − 32) × 0.556$$

When recording air temperature, the degree symbol is placed before the C or F and is read *degree Celsius* (°C) or *degree Fahrenheit* (°F)—as in "yesterday was 20°C and today is 25°C." However, when expressing the difference between two temperatures, dealing with lapse rates, temperature ranges, or units of temperature, the degree symbol is placed after the C or F and is read *Celsius degree* (C°) or *Fahrenheit degree* (F°)—as in "the high temperature yesterday was 20°C, which is 5 C° cooler than today." Another example is that ice melts at 0°C and boils at 100°C, which means there are 100 C° between melting and boiling.

The formulas to use to convert between a number of Celsius and Fahrenheit degrees are:

To convert C° to F°:

$$F° = C° × 1.8 \quad \text{or} \quad F° = 1.8 \, C°$$

To convert F° to C°:

$$C° = F° ÷ 1.8 \quad \text{or} \quad C° = F° × 0.556$$

1. Using the given conversion formulas, complete the following temperature conversions. The first answer is provided for you in bracketed italics.

   25°C _77° F*_   30°F _____   −18°C _____   −9°F _____

   4°C _____   92°F _____   57°C _____   57°F _____

   134°F _____   59°F _____   0°C _____   15°C _____

   * {°F = [(25° × 1.8) + 32°] = (45° + 32°) = 77° F}

2. Using the given conversion formulas, complete the following degree conversions.

   5 F° _____   25 C° _____   15 F° _____   18 C° _____

   6 C° _____   90 F° _____   27 C° _____   32 F° _____

3. What temperature is it as you work on this lab exercise?

   a) Outdoor temperature? _____°C   _____°F

   b) Indoor temperature? _____°C   _____°F

4. Using a physical geography text, an atlas, an encyclopedia, or the Internet, answer the following. Be sure to list temperature record, date, and place.

   a) Highest natural temperature recorded on Earth (where and value)?

   b) Lowest natural temperature recorded on Earth (where and value)?

   c) Lowest natural temperature recorded for the Northern Hemisphere?

   d) Highest natural temperature recorded for the Southern Hemisphere?

SECTION 2

The Temperatures We Feel

Our perception of temperature is described by the terms **apparent temperature** or *sensible temperature*. The combination of water vapor content, wind speed, and air temperature affect each individual's sense of comfort. High temperatures, high humidity, and low winds produce the most heat discomfort, whereas low humidity and strong winds enhance cooling sensation and effect (due to the increased evaporation of moisture from the skin). The wind chill index (given by the National Weather Service) is important to those who experience winters with freezing temperatures. The National Weather Service and the Meteorological Services of Canada (http://www.nws.noaa.gov/os/ cold/wind_chill.shtml) revised the wind chill formula and standard assumptions and introduced a new wind-chill chart during the 2001–2002 winter season. The new Wind Chill Temperature (WCT) Index is an effort to improve the accuracy of heat loss calculations.

A wind-chill chart of estimated values is presented in **Figure 7.1**, in both metric and English units. The **wind chill factor** indicates the enhanced rate at which body heat is lost to the air. As wind speeds increase, heat loss from the skin increases. For example, if the air temperature is −1°C (30°F) and the wind is blowing at 32 kmph (20 mph), skin temperatures will be −8°C (17°F).

Actual Air Temperature in °C (°F)

Calm	4° (40°)	−1° (30°)	−7° (20°)	−12° (10°)	−18° (0°)	−23° (−10°)	−29° (−20°)	−34° (−30°)	−40° (−40°)
8 (5)	2° (36°)	−4° (25°)	−11° (13°)	−17° (1°)	−24° (−11°)	−30° (−22°)	−37° (−34°)	−43° (−46°)	−49° (−57°)
16 (10)	1° (34°)	−6° (21°)	−13° (9°)	−20° (−4°)	−27° (−16°)	−33° (−28°)	−41° (−41°)	−47° (−53°)	−54° (−66°)
24 (15)	0° (32°)	−7° (19°)	−14° (6°)	−22° (−7°)	−28° (−19°)	−36° (−32°)	−43° (−45°)	−50° (−58°)	−57° (−71°)
32 (20)	−1° (30°)	−8° (17°)	−16° (4°)	−23° (−9°)	−30° (−22°)	−37° (−35°)	−44° (−48°)	−52° (−61°)	−59° (−74°)
40 (25)	−2° (29°)	−9° (16°)	−16° (3°)	−24° (−11°)	−31° (−24°)	−38° (−37°)	−46° (−51°)	−53° (−64°)	−61° (−78°)
48 (30)	−2° (28°)	−9° (15°)	−17° (−1°)	−24° (−12°)	−32° (−26°)	−39° (−39°)	−47° (−53°)	−55° (−67°)	−62° (−80°)
56 (35)	−2° (28°)	−10° (14°)	−18° (0°)	−26° (−14°)	−33° (−27°)	−41° (−41°)	−48° (−55°)	−56° (−69°)	−63° (−82°)
64 (40)	−3° (27°)	−11° (13°)	−18° (−1°)	−26° (−15°)	−34° (−29°)	−42° (−43°)	−49° (−57°)	−57° (−71°)	−64° (−84°)
72 (45)	−3° (26°)	−11° (12°)	−19° (−2°)	−27° (−16°)	−34° (−30°)	−42° (−44°)	−50° (−58°)	−58° (−72°)	−66° (−86°)
80 (50)	−3° (26°)	−11° (12°)	−19° (−3°)	−27° (−17°)	−35° (−31°)	−43° (−45°)	−51° (−60°)	−59° (−74°)	−67° (−88°)

Wind speed, kmph (mph)

Frostbite times: ☐ 30 min. ◼ 10 min. ◼ 5 min.

▲ **Figure 7.1** Wind chill chart for various temperatures and wind velocities

1. Use the wind chill chart in Figure 7.1 to determine the wind chill temperature for each of the following examples:

°C   (°F)

    a) Wind speed: 24 kmph, air temperature: -34°C $=$ wind-chill temp: _____

    b) Wind speed: 48 kmph, air temperature: -7°C $=$ wind-chill temp: _____

    c) Wind speed: 8 kmph, air temperature: $+4$°C $=$ wind-chill temp: _____

    d) Wind speed: 56 kmph, air temperature: -23°C $=$ wind-chill temp: _____

2. Competitive downhill ski racers are subjected to severe wind chill, and so are average skiers and snow-boarders, to a lesser degree. Assuming a downhill racer is going 80 kmph (50 mph), which is coasting on some runs, and the air temperature is -18°C (0°F), what is the wind chill the skier is feeling on any exposed skin?

What is the skier's time to experience frostbite, given these conditions?

The **heat index** (HI) indicates the human body's reaction to air temperature and water vapor. The water vapor in air is expressed as relative humidity, a concept presented in Lab Exercise 12. For now, we can simply assume that the amount of water vapor in the air affects the evaporation rate of perspiration from the skin because the more water vapor in the air (the higher the humidity), the less water from perspiration the air can absorb through evaporation, thus reducing natural evaporative cooling. The heat index, **Figure 7.2**, is an abbreviated version of the heat index used by the NWS and now is included in its daily weather summaries during appropriate months. The table beneath the graph describes the effects of heat index categories on higher-risk groups. A combination of high temperature and high humidity can severely reduce the body's natural ability to regulate internal temperature. The NWS provides a heat index forecast warning map (see http://www.nws.noaa.gov/om/heat/index.shtml); a heat index calculator is provided at http://www.crh.noaa.gov/jkl/?n=heat_index_calculator.

3. Use the heat index chart in Figure 7.2 to determine the heat index temperature for each of the following examples:

°C   (°F)

    a) Air temperature: 37.8°C, relative humidity 5% $=$ heat index temp: _____

    b) Air temperature: 32.2°C, relative humidity 80% $=$ heat index temp: _____

    c) Air temperature: 32.2°C, relative humidity 90% $=$ heat index temp: _____

    d) Air temperature: 43.3°C, relative humidity 10% $=$ heat index temp: _____

4. List the temperatures and heat index categories you would experience if the temperature stayed a constant 35°C (95°F) but the relative humidity dropped from 90% down to 10%. List the temperatures and heat index categories at each drop of 10% relative humidity (90%, 80%, 70%...).

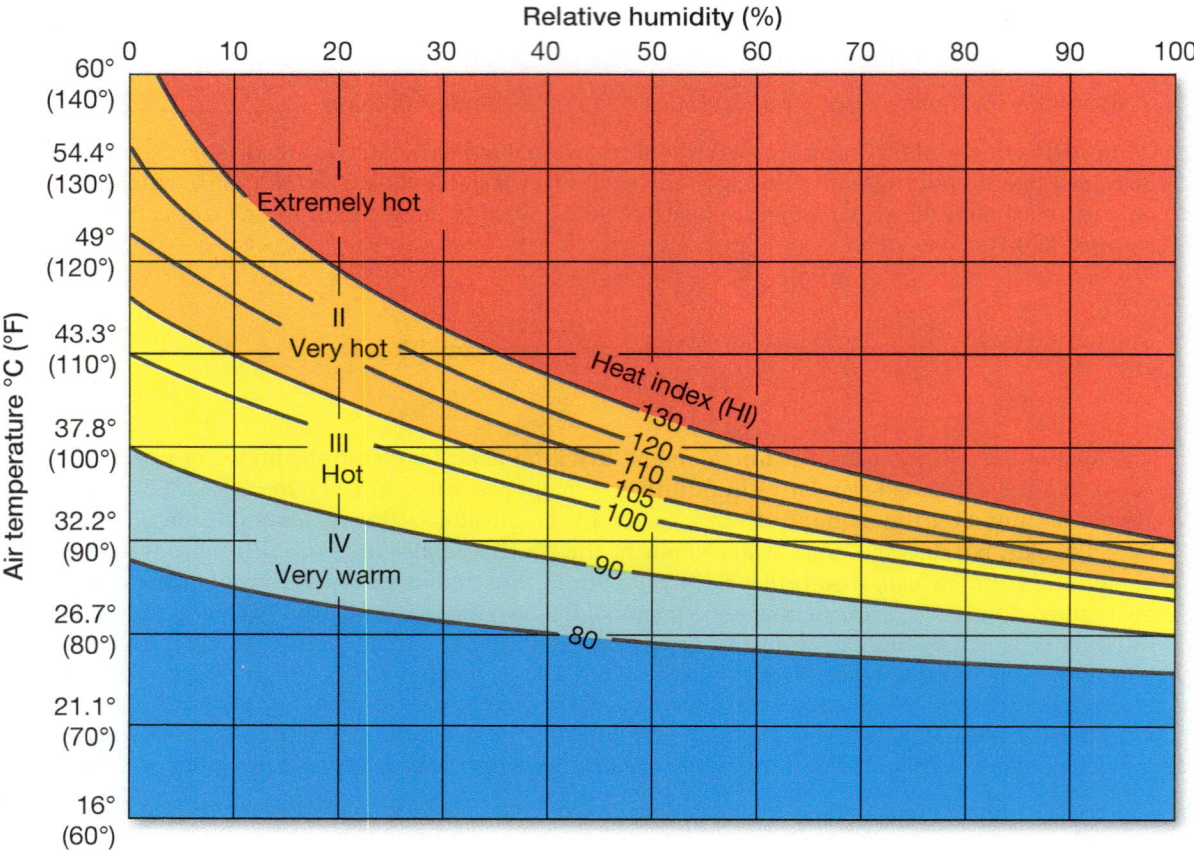

Level of concern	Category	Heat Index Apparent Temperature	General Effect of Heat Index on People in High-Risk Groups
Extreme danger	I	54°C (130°F) or higher	Heat/sunstroke highly likely with continued exposure
Danger	II	41° – 54°C (105° – 130°F)	Sunstroke, heat cramps, or heat exhaustion likely and heatstroke possible with prolonged exposure and/or physical activity
Extreme caution	III	32° – 41°C (90° – 105°F)	Sunstroke, heat cramps, and heat exhaustion possible with prolonged exposure and/or physical activity
Caution	IV	27° – 32°C (80° – 90°F)	Fatigue possible with prolonged exposure and/or physical activity

▲ **Figure 7.2** Heat index for various temperatures and relative humidity levels.
The Heat Index (HI) numbers indicate the Fahrenheit temperature it would feel like, with the given temperature and humidity conditions.

SECTION 2

Latitude and Parallels

Adapting from the Babylonians, Ptolemy divided the circle into 360 *degrees* (360°), with each degree subdivided into 60 *minutes* (60′), and each minute further subdivided into 60 *seconds* (60″). This method of dividing degrees into minutes and seconds is often referred to as DMS coordinates. Geographic Information Systems, GPS, and other computer-based systems often use decimal degrees or DD coordinates. In this method, each degree is divided into a more familiar base 10 system. We'll now examine each of these grid coordinate elements.

Latitude is an angular or arc distance north or south of the *equator* (the line running east to west halfway between the poles), measured from the center of Earth (**Figure 1.2a**). On a map or globe,

the lines designating these angles of latitude run east and west, parallel to the equator, whose latitude is 00.00°.

The North Pole is 90.00° north latitude, and the South Pole is 90.00° south latitude. "Lower latitudes" are those nearer the equator, whereas "higher latitudes" refer to those nearer the poles.

A line connecting all points at the same latitudinal angle is called a **parallel**. Only one parallel—the equatorial parallel—is a great circle; all others are small circles, diminishing in circumference north and south toward the poles, which are merely points. As shown in Figures 1.2a and 1.2b, *latitude* is the name of the angle (49° north latitude), *parallel* names the line (49th parallel), and both indicate arc distance north of the equator.

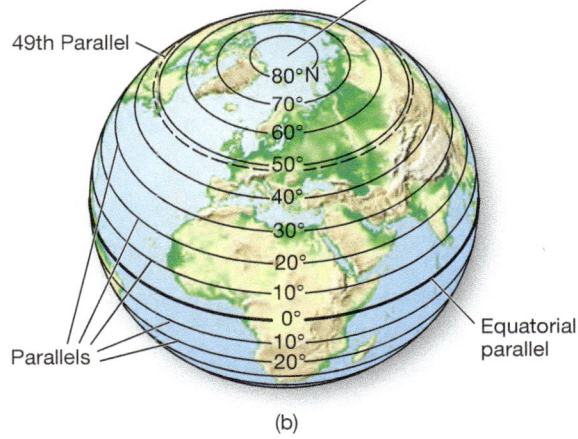

▲ **Figure 1.2 a and b** Latitude and parallels

1. Parallels have often been used to demarcate political boundaries. The 49th parallel north forms a portion of the border between which two countries?

2. Parallels made famous by wars in the last century include the parallel dividing the two Koreas and the parallel that divided Vietnam until 1975. (The border was approximately 80 km [50 mi] north of the city of Hue.) What are these two parallels?

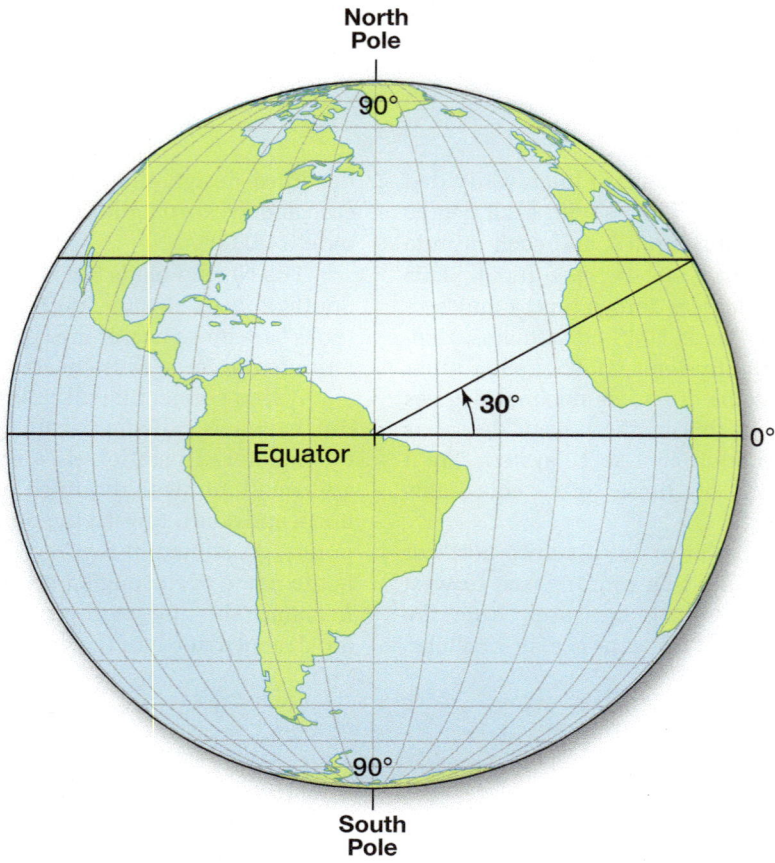

▲ **Figure 1.3** Measuring latitudes on Earth

3. **Figure 1.3** is a half-Earth section through the poles. An angle of 30° has been measured north of the equator, with the 30th parallel drawn on the globe.

   a) Following this example, use your protractor to measure an angle of 23.5° north of the equator and draw and label the corresponding parallel, which is known as the Tropic of Cancer.

   b) Measure another angle 66.5° north of the equator, and draw and label the parallel known as the Arctic Circle.

   c) Repeat this in the Southern Hemisphere, drawing and labeling the Tropic of Capricorn (23.5° south) and the Antarctic Circle (66.5° south). We will be working with these parallels while studying Earth–Sun relationships in Lab Exercise 5.

   d) Finally, measure, draw, and label one last parallel to show your own latitudinal location.

4. Using a globe and your atlas or other world map, locate three cities that are located at approximately the 23rd parallel in the Northern Hemisphere; note their location in degrees (and minutes, if your map is detailed enough to estimate minutes). Use the globe first, then refer to the atlas maps to better determine specific latitudes. Be sure and list their country names as well.

   City and Country Name   **Longitude (degrees and minutes, if possible)**

   a) _____   _____

   b) _____   _____

   c) _____   _____

5. List the names and longitudes of three cities that are located at approximately your latitude.

City and Country Name	Longitude
a) _____	_____
b) _____	_____
c) _____	_____

SECTION 3

Longitude and Meridians

Longitude is an angular or **arc distance** east or west of a point on Earth's surface, measured from the center of Earth (**Figure 1.4a**). On a map or globe, the lines designating these angles of longitude run north and south at right angles (90°) to the equator and to all parallels. A line connecting all points at the same longitude is a **meridian** (**Figure 1.4b**). Every meridian is one-half of a great circle that passes through the poles. As shown in Figures 1.4a and 1.4b, longitude is the name of the angle, meridian names the line, and both indicate arc distance east or west of the **prime meridian**.

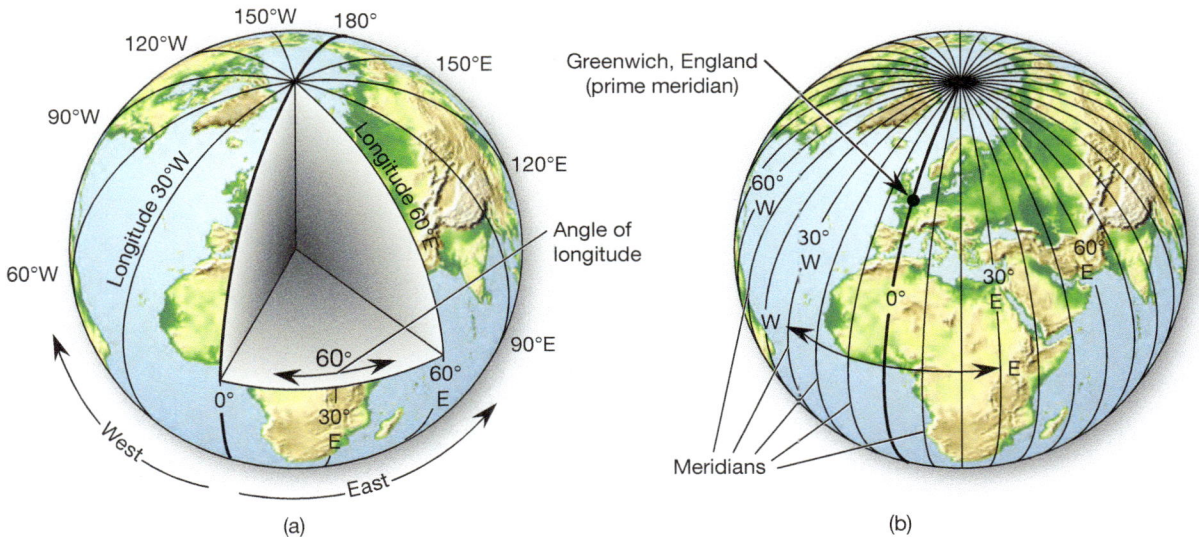

▲ **Figure 1.4 a and b** Longitude and meridians

1. On a political globe or world map follow the International Date Line across the Pacific Ocean. Why do you think the International Date Line is not straight but zigs and zags?

2. Examine an atlas or a political globe and in the spaces marked a through h list the provinces and states through which the 100th meridian in the Western Hemisphere passes—north to south. The first answer is provided for you in bracketed italics.

a)	*[Nunavut, Canada]*	e)	_____	
b)	_____	f)	_____	
c)	_____	g)	_____	
d)	_____	h)	_____	

3. **Figure 1.5** is a view of Earth from directly above the North Pole; the equator is the full circumference around the edge. A line has been drawn from the North Pole to the equator and labeled 0°, representing the prime meridian. Earth's prime meridian through Greenwich, England, was not generally agreed to by most nations until 1884. To the right of 0° on the diagram is the *Eastern Hemisphere*, and to the left of 0° is the *Western Hemisphere*.

   a) Label both the Eastern Hemisphere and the Western Hemisphere.

   b) Extend another line from the North Pole to the other side of Earth, opposite the prime meridian, and label it 180°. You now have marked the line that is the International Date Line, which extends from North to South Poles on the opposite side of Earth from the prime meridian.

   c) Using your protractor, measure, draw, and label the meridians that are 100° east and 60° west of the Greenwich meridian.

   d) Finally, locate, draw, and label the meridian that marks your present longitude.

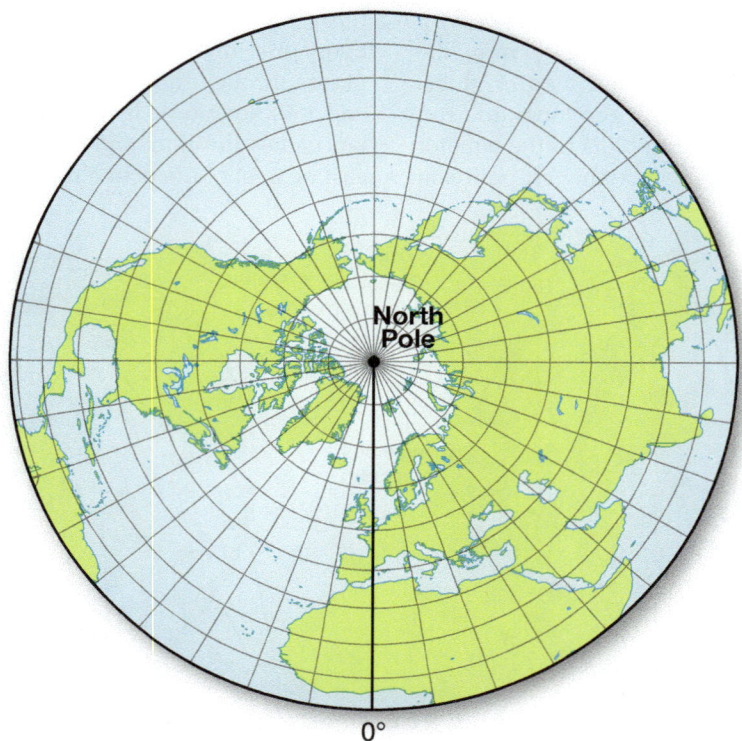

▲ **Figure 1.5** Measuring longitudes on Earth

The 98th meridian is roughly the location of the 51-cm (20-in.) isohyet (a line connecting points of equal precipitation), with wetter conditions to the east and drier conditions to the west. In North America, tallgrass prairies once rose to heights of 2 m (6.5 ft) and extended westward across the Great Plains to about the 98th meridian, with short-grass prairies farther west. The deep sod that formed beneath these tall grasslands posed problems for the first settlers, as did the climate. The self-scouring steel plow, introduced in 1837 by John Deere, allowed the interlaced grass sod to be broken apart, freeing the soils for agriculture. Keep this information about the region around the 100th meridian, and close to the 98th, in mind as a "location reference" later on when studying precipitation and vegetation.

Mastering Geography

Looking for additional lab prep and review materials? Visit the Study Area in Mastering Geography to enhance your geographic literacy and spatial reasoning skills by accessing MapMaster™ interactive maps, videos, *Mobile Field Trips*, *Project Condor* Quadcopter Videos, *In the News* articles, flashcards, web links, self-study quizzes, and other resources. Contact your professor to find out how you can access Mastering Geography.

Name: _____   Laboratory Section: _____

Date: _____   Score/Grade: _____

Video
Exercise 2
Pre-Lab Video

http://goo.gl/lbWTVC

Scan to view the
Pre-Lab video

LAB EXERCISE

The Geographic Grid and Time

In the previous exercise we studied latitude and longitude. In this exercise we will examine latitude and longitude together as they make the geographic grid that we use to locate places. As you complete your *Geography I.D.* in the Preface, you will determine the time zone you live in. A knowledge of longitude and Earth's rotation forms the basis of standard time and the world system of Coordinated Universal Time (UTC). We examine both this essential geographic location grid and time in this exercise. Lab Exercise 2 features three sections.

Key Terms and Concepts

antipode
Coordinated Universal Time (UTC)
daylight saving time

geographic grid
Greenwich Mean Time (GMT)

International Date Line
local Sun time

KEY**LEARNING**concepts

After completion of this lab you should be able to:

- **Use** latitude and longitude to *locate* places on Earth's surface.
- **Relate** the time where you are with world standard time and *calculate* differences in standard and Sun time.
- **Contrast** the prime meridian with the International Date Line.

Materials/Sources Needed

protractor
world atlas

Lab Exercise and Activities

SECTION 1

Earth's Geographic Grid

Latitude and longitude form Earth's **geographic grid**. Every location on our planet is described with this Cartesian coordinate system—where location is described on a plane by two intersecting lines. The *Geography I.D.* asks you to identify the general coordinates of your hometown and college campus. **Antipodes** are points on the globe that are diametrically opposite each other and would be at either end of a line drawn through the center of Earth—such as the North Pole and South Pole. Determine the following geographic grid coordinates and cities from an atlas, wall map, or large political world globe. While wall maps and globes will help you see points in relation to the rest of the world, you should use regional maps in an atlas to be more precise in your readings.

1. Locate and give the geographic coordinates for the following cities (to a tenth of a degree if your atlas maps are detailed enough) or identify the cities from the given coordinates. The answers to a) and e) are provided for you in bracketed italics. Once you have identified the cities and found the coordinates, plot the coordinates in items 1 (a) through (h) above on the map grid in Figure 2.1, and label the city names.

City	Latitude and Longitude
a) Greenwich, London, England	*[51.5°N 0°]*
b) Rio de Janeiro, Brazil	
c) Sydney, Australia	
d) Your state's/province's capital city	
e) *[Tokyo, Japan]*	35.7°N 139.7°E
f)	8.8°S 13.2°E
g)	21.3°N, 157.8°W
h)	54.8°S, 68.3°W

2. If you were halfway between the equator and the South Pole and one-quarter of the way around Earth to the west of the prime meridian, what would be your latitude and longitude?

3. You are at 10°N and 30°E; you move to a new location that is 25° south and 40° west of your present location. What is your new latitudinal/longitudinal position?

4. You are at 20°S and 165°E; you move to a new location that is 45° north and 50° east from your present location. What is your new latitudinal/longitudinal position?

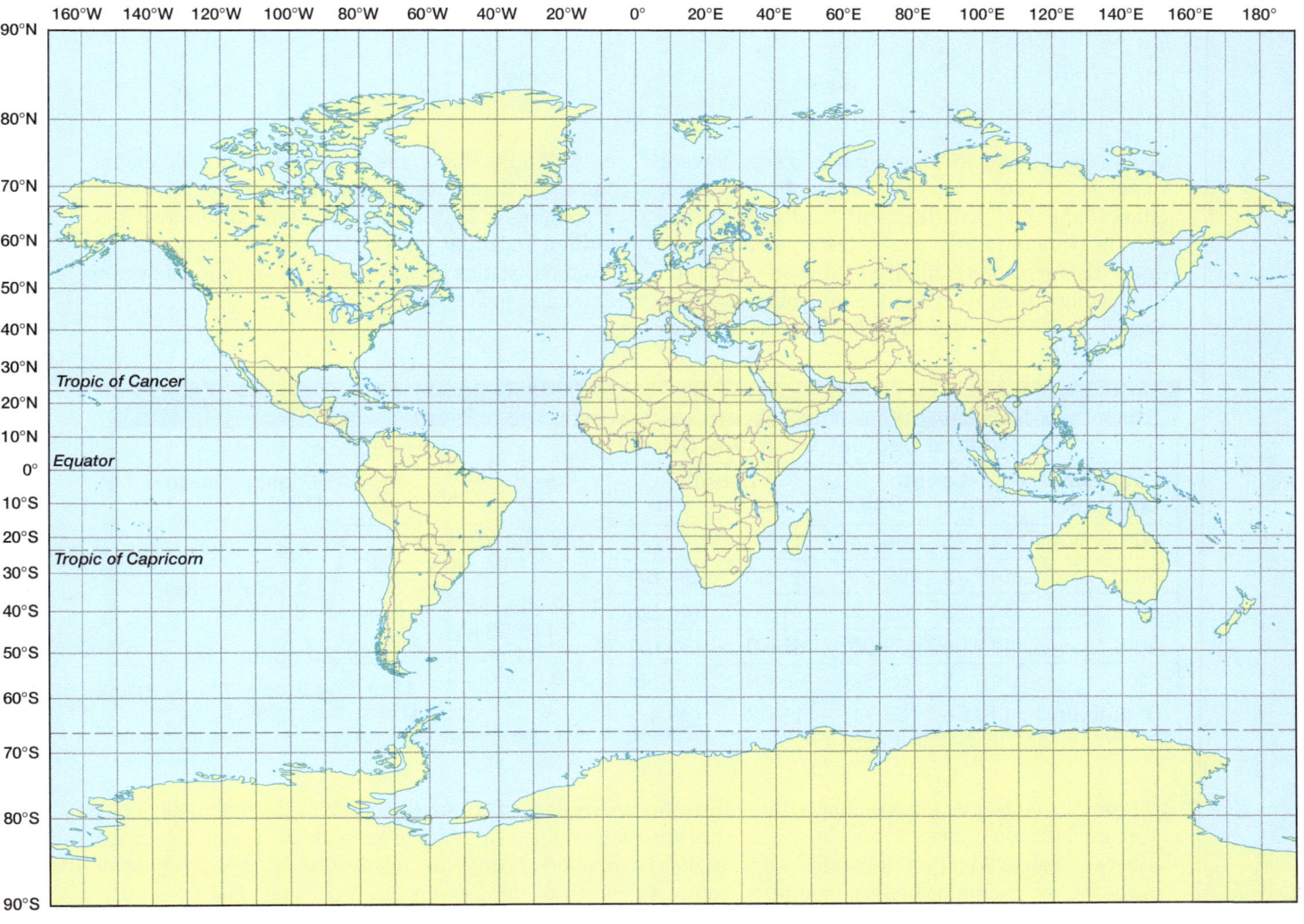

▲ **Figure 2.1** Plotting coordinates

The antipode is the point on the opposite side of Earth from another point. You may have heard that if you dig straight down, you'll eventually reach China. Setting aside the difficulties of digging through the molten iron outer core of Earth, you wouldn't end up in China if you started digging from the United States.

To find the antipode of a location, you'll need to find both the latitude and longitude of the antipode. To find the antipodean latitude, convert the location's latitude to the opposite hemisphere. If the latitude is 50°N, it's antipodean latitude is 50°S. With longitude, the antipode is always 180° away. To find the antipodean latitude, simply subtract the location's latitude from 180° and change it to the other hemisphere. To find the antipodean longitude of 120°W, subtract 120° from 180°, and change it to the Eastern Hemisphere to find 60°E.

5. What is the antipode of your current location?

6. If you wanted to dig through the center of the Earth and come up in Beijing, China, where should you start digging?

SECTION 2

Latitude and Longitude Values

Because meridians of longitude converge toward the poles, the actual distance on the ground (linear distance) spanned by a degree of longitude is greatest at the equator (where meridians separate to their widest distance apart) and diminishes to zero at the poles (where they converge). The linear distance covered by a degree of latitude, however, varies only slightly, due to the oblateness of Earth's shape. **Table 2.1** compares the linear or physical distance of a degree of latitude and longitude at selected latitudinal locations. It also shows the similarity of ground distance for a degree of latitude and a degree of longitude at the equator.

TABLE 2.1 Physical distances represented by degrees of latitude and longitude

Latitudinal Location	Latitude Degree Length km	Latitude Degree Length (mi)	Longitude Degree Length km	Longitude Degree Length (mi)
90° (poles)	111.70	(69.41)	0	(0)
60°	111.42	(69.23)	55.80	(34.67)
50°	111.23	(69.12)	71.70	(44.55)
40°	111.04	(69.00)	85.40	(53.07)
30°	110.86	(68.89)	96.49	(59.96)
0° (equator)	110.58	(68.71)	111.32	(69.17)

North Pole
1° of latitude = 111.70 km
1° of longitude = 0 km
1° of latitude = 111.04 km   40°
1° of longitude = 85.40 km
1° of latitude = 110.58 km (1 arc of a meridian)
Equator
1° of longitude = 111.32 km (1 arc of a parallel)

1. From the table, you can see that latitude lines are evenly spaced, approximately 111 km (69 miles) apart at any latitude. Using these values as the linear distance separating each degree of latitude, the distance between any given pair of parallels can be calculated. (*Note: Locations must be due north-south of each other.*) For example, Denver is approximately 40° north of the equator (arc distance). The linear distance between Denver and the equator can be calculated as follows:

   40° N to 0° = 40° × 111 km/1° = 4440 km

   or

   40° N to 0° = 40° × 69 miles/1° = 2760 miles

   Using these same values for a degree of latitude and an atlas for city location, calculate the linear distance in kilometers and miles between the following sets of points (along a meridian):

   a) Mumbai, India, and the equator _____

   b) Miami, Florida, and 10° south latitude _____

   c) Edinburgh, Scotland, and 5° north _____

   d) Your location and the equator _____

2. The table also shows that the linear distance separating each 1° of longitude decreases toward the poles. For example, at 30° latitude each degree of longitude is separated by slightly more than 96 km (nearly 60 miles), and at 60° latitude, the linear distance is reduced to approximately half that at the equator. For each of the following latitudes, determine the linear distance in kilometers and in miles for 15° of longitudinal arc (along a parallel): The first answer is provided for you in bracketed italics.

	km	**miles**
a) 30° latitude:	[15° × 96.49 km = 1447 km]	[15° × 59.96 mi = 899 mi]
b) 40° latitude:		
c) 50° latitude:		
d) 60° latitude:		

3. Again using Table 2.1, what is the linear distance in kilometers and miles along the parallel at your latitude from your location to the prime meridian?

4. What is the approximate linear distance of the following angular distances, at your present latitude?

	km	**miles**
One degree		
One minute of longitude		
One second of longitude		
A tenth of a degree		
One hundredth of a degree		

Precision and accuracy are two related concepts regarding measurements. *Precision* refers to the number of significant digits in a measurement, while *accuracy* refers to how close a measurement is to the actual. Giving a distance measurement as 2 km is less precise than giving it as 2.0 km or 2.00 km. If you were going to meet a friend in the Lake Nakuru National Park in Kenya, and your friend's location was 0°, 36°E, how many square kilometers would you have to look through? Your friend's location is precise to a full degree so you would have to look in an area 1° by 1°. At this location, 1° of both longitude and latitude are approximately 111 km (69 mi), so you would have to look for your friend in 12,321 km^2 (4761 mi^2). If your friend's location was given 10 times as precisely, as 0.0°, 36.0°E, you would have to look in an area 1/10th of a degree by 1/10th of a degree, which would be 11.1 km by 11.1 km, or 123.2 km^2.

5. How large an area would you have to look through if your friend's location were given as 0.00°, 36.00°E?

6. Write your friend's location with sufficient precision so that you would only have to look in an area 111 m by 111 m.

SECTION 3

Time, Time Zones, and the International Date Line

Today we take for granted international standard time zones and an agreed-upon prime meridian. If you live in Oklahoma City and it is 3:00 P.M., you know that it is 4:00 P.M. in Baltimore and Toronto, 2:00 P.M. in Salt Lake City, and 1:00 P.M. in Seattle and Los Angeles. You also probably realize that it is 9:00 P.M. in London and midnight in Riyadh, Saudi Arabia. (The designation A.M. is for *ante meridiem*, for "before midday/noon," whereas P.M. is for *post meridiem*, meaning "after midday/noon.") Coordination of international trade, airline schedules, business activities, and daily living depends on a common time system.

In 1884 at the International Meridian Conference in Washington, DC, the prime meridian was set as the official standard for the world time zone system—**Greenwich Mean Time (GMT)** (see http://wwp.greenwich-meantime.com/). Coordinated Universal Time For decades, GMT was determined using the Royal Observatory's astronomical clocks and was the world's standard for accuracy. In 1972, the **Coordinated Universal Time (UTC)** time-signal system replaced GMT and became the legal reference for official time in all countries. UTC is based on average time calculations from atomic clocks collected worldwide. You might still see official UTC referred to as GMT or Zulu time.

Because Earth revolves 360° every 24 hours, or 15° per hour (360° ÷ 24 = 15°), a time zone of 1 hour is established for each 15° of longitude. Each time zone theoretically covers 7.5° on either side of a controlling (standard) meridian (0°, 15°, 30°, 45°, 60°, 75°, 90°, 105°, 120°, etc.) and represents 1 hour. Traveling eastward, one would set the clock 1 hour later, and going westward, 1 hour earlier for each time zone.

Local Sun time, or solar time, is based on the actual longitudinal arc distance (in degrees, minutes, and seconds) between a location and the

▲ **Figure 2.2** Modern international standard time zones.
The numbers at the bottom indicate how many hours each zone is ahead (plus sign) or behind (minus sign) Greenwich Mean Time, today known as Coordinated Universal Time (UTC) or Zulu Time.

prime meridian. There could be as much as 30 minutes' difference in time (or more, depending on the actual time zone boundaries) between local Sun time and standard time.

Opposite the globe from the prime meridian is the **International Date Line**, the starting point for each new day. Since the Date Line is the standard (controlling) meridian for a time zone, no change in time occurs when crossing the line, only the date. When traveling westward, one will set the calendar 1 day ahead, and when crossing the line heading eastward, the calendar will be reset 1 day back.

1. From the map of global time zones in **Figure 2.2**, determine the present time in the following cities: (For your time, use the starting time of the lab.)

   Moscow _____   Los Angeles _____

   London _____   Honolulu _____

   Chicago _____   Mumbai _____

2. You may not always have a time zone map available, but by remembering the relationship of 1 hour for every 15° of longitude, you can easily calculate the difference in time between places. Indicating and using the standard meridians to determine time zones, solve the following problems. The first answer is provided for you in bracketed italics. Show your work:

   a) If it is 3 A.M. Wednesday in Vladivostok, Russia (132°E), what day and time is it in Moscow (37°E)? [*The controlling meridian for Moscow is 105° away from Vladivostok's controlling meridian of 135°E (135° − 30° = 105° difference). Since Earth rotates 15° per hour, Moscow is 7 hours earlier than Vladivostok (105° difference / 15° rotation per hour = 7 hours time difference), therefore if it is 3 A.M. Wednesday in Vladivostock, it is 8 P.M. Tuesday in Moscow.*]

   b) If it is 7:30 P.M. Thursday in Winnipeg, Manitoba, Canada (97°W), what day and time is it in Harare, Zimbabwe (31°E)?

   c) If you depart from San Francisco International Airport at 10:00 P.M. on Tuesday, what day and time will you arrive in Auckland, New Zealand (175°E), assuming a flight time of 14 hours?

3. If there is a difference of 15° of longitude for each hour of time, how much difference in time is there for 1° of longitude? for 1′ of longitude?

4. Which of the following standard (controlling) meridians is the standard meridian for your time zone? 75°—Eastern; 90°—Central; 105°—Mountain; 120°—Pacific; 135°—Alaska; other.

   How many degrees of longitude separate you from this standard controlling meridian?

   Calculate the difference between standard and Sun time using the answer you determined in no. 3 above. How many minutes ahead or behind your standard meridian are you?

Knowing your standard meridian is useful because it will tell you how many minutes off (fast or slow) your watch is in your time zone from the actual position of the Sun in the sky.

Just as you can use longitudinal distance to calculate time, you can also use time difference to calculate your longitude. This is the method that was often used to determine longitudinal location when sailing (if there wasn't a GPS unit on board!). Again, the relationship between time and longitude ($15° = 1$ hour of time) is the key. If you know the difference in time between two places, and you know the longitude of one place, you can calculate the longitude of the other. For example: Your watch (kept on your hometown standard time) reads 10:30 A.M., and you see that a clock in the town you are visiting reads 7:30 P.M. Your home near St. Paul, Minnesota, uses 90°W as its standard meridian. Therefore:

- Time difference between 10:30 A.M. and 7:30 P.M. = 9 hours.
- 9 hours × 15°/hour = 135° of longitude separating the two locations.
- Since you are at a place later in time than home, you must be east of home.
- 135° east of your hometown at 90°W would be 45°E longitude.

5. Assume the time on your watch, showing local standard time, is 4:15 P.M. A chronometer reads 2:15 A.M. What is your longitude?

6. John Harrison's chronometer (a clock giving Coordinated Universal Time) lost 5 seconds during the 81-day voyage from England to Jamaica. Given that Earth rotates through 15° in 1 hour, how many degrees of longitude would the ship be off, with an error of 5 seconds? How many kilometers and miles would that be, assuming 111 km per degree of longitude?

Daylight saving time is the practice of setting time ahead 1 hour in the spring and back 1 hour in the fall in the Northern Hemisphere. Clocks in the United States were left advanced an hour continuously from 1942 to 1945 during World War II and again from January 1974 to October 1975, in order to conserve energy during an oil embargo. Daylight saving time increased in length in the United States and Canada in 1986. Under the Energy Policy Act of 2005, daylight saving time begins 3 weeks earlier than previously, at 2 A.M. on the second Sunday in March. DST has been extended by 1 week and now lasts until 2 A.M. on the first Sunday of November.

7. Does your community adopt daylight saving time? What are the dates for adjusting clocks in the spring and fall?

8. What time does your physical geography lab start
   a) according to standard time? _____
   b) according to daylight saving time? _____
   c) in UTC? _____
      (24-hour clock time in Greenwich, England, e.g., 3:00 P.M. = 15:00 hours)

Mastering Geography

Looking for additional lab prep and review materials? Visit the Study Area in Mastering Geography to enhance your geographic literacy and spatial reasoning skills by accessing MapMaster™ interactive maps, videos, *Mobile Field Trips*, *Project Condor* Quadcopter Videos, *In the News* articles, flashcards, web links, self-study quizzes, and other resources. Contact your professor to find out how you can access Mastering Geography.

Video
Exercise 3
Pre-Lab Video

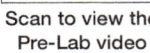
http://goo.gl/L5b275

Scan to view the
Pre-Lab video

LAB EXERCISE

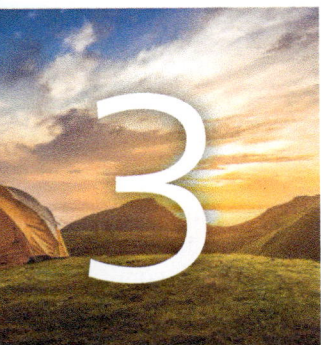

Directions and Compass Readings

Earth's coordinate grid system was covered in Lab Exercise 2, so you are familiar with using latitude and longitude to find a point's absolute location on Earth's surface. There are times when relative direction between two places is desired—for example if you want to navigate from one point to another. There are two systems used to indicate relative direction on maps: compass points and azimuths.

Understanding and being able to use these direction-finding methods is important, whether you are reading a street/road map or are interpreting specialized maps such as topographic maps, which are discussed and used in later exercises. These systems are all measurements of horizontal direction, as if you were in the air looking down at Earth from above.

Compasses are useful for finding directions, but corrections must be made for deviation from true or geographic north. Lab Exercise 3 has four sections.

Key Terms and Concepts

azimuths
cardinal points
compass points
geographic north
geographic south

grid north
isogonal map
isogons
magnetic compass
magnetic declination

magnetic north
topographic map
true north
true south

KEY**LEARNING**concepts

After completion of this lab, you should be able to:

- **Describe** the method of indicating direction using compass readings and *apply* that method to compass directions.
- **Define** magnetic declination and *determine* differences in "north" readings on a map.
- **Explain** any difference between a local topographic map and the isogonal map relative to magnetic declination.

Materials/Sources Needed

color pencils
compass
local topographic map

protractor
ruler

Lab Exercise and Activities

SECTION 1

Compass Points

The directional system familiar to most people is that of **compass points**. The symbol below is a *compass rose*. Cartographers often included elaborate compass roses as a decorative, as well as a functional, part of antique maps. The compass rose has four **cardinal points**—in a clockwise direction—north, east, south, and west, separated by four intermediate points: NE, SE, SW, NW. These are split into 16 and then, sometimes, again into 32 compass points. You may not be familiar with compass points such as NNE and ENE. These points are used when greater precision is needed. NNE (read north northeast) is north of NE, while ENE (read east northeast) is east of NE.

1. Label 16 compass points on **Figure 3.1**, using color pencils on the compass rose to distinguish the categories of division. (Include a legend for your color scheme.) In each division, the points are listed clockwise.

N	NE	NNE
E	SE	ENE
S	NW	ESE
W	SW	SSE
		SSW
		WSW
		WNW
		NNW

▲ **Figure 3.1** Compass points

SECTION 2

Azimuths

While useful, compass points are not extremely accurate, and most often the direction between two points does not fall exactly on one of the 32 points. Degrees of an arc are used for more precise directions.

Azimuths are read from the north in a clockwise direction, from 0° to 360° (with 0° and 360° being the same point—north). For further precision, the degrees can be broken down into minutes (′).

1. **Figure 3.2** has a few azimuths drawn and labeled as examples. Using your protractor, determine the azimuth readings for *A* and *B*, and label the value for each on the diagram.

2. Measure, draw, and label the following azimuths on the diagram: 230°, 78°, 145°.

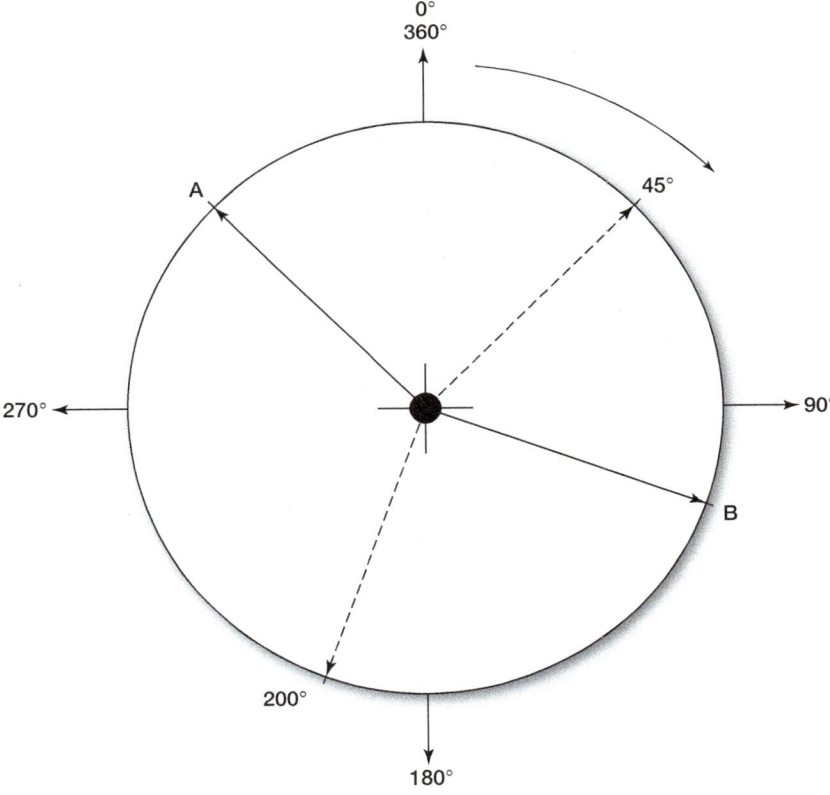

▲ **Figure 3.2** Azimuths

3. Using Figure 3.1, Figure 3.2, and your protractor, answer the following questions.

Which azimuth is found at the following compass points: NE ____[45°]____ SW _____

W _____ ESE _____ NNW _____

Which compass point is found at the following azimuths: 135° ____[SE]____ 247.5° _____

180° _____ 315° _____ 225° _____

Compass Declination

Earth rotates around an axis that is marked by the **geographic North** and **South Poles**, also referred to as **true north** and **true south**. Meridians of longitude converge at these geographic points.

A **magnetic compass**, an instrument used to determine direction, points to the **magnetic North Pole**, which does not coincide with true north and the geographic pole. The difference between the north arrow on a magnetic compass and true north is the **magnetic declination**. A compass needle that points west of true north indicates a *west declination*, and a compass needle that points east of true north indicates an *east declination*. When using a magnetic compass, you must make adjustments for the difference between magnetic north and true north.

Isogons, lines connecting points of equal magnetic declination, are shown on **isogonal maps**, such as in **Figure 3.3**. The magnetic poles change over time, slowly migrating, so that adjustments for these magnetic pole changes need to be considered if map navigation is to be up-to-date.

▲ **Figure 3.3** Magnetic field declination, 2020 (in degrees)

1. Using the isogonal map in Figure 3.3 and your atlas, in what city would you have to stand during 2020 to get a compass reading with zero magnetic declination (where magnetic north aligns with true north, or an agonic line). Mark the city with a dot and a label on the map.

2. Using your atlas, turn to the pages featuring the polar regions and find the latitude/longitude coordinates for the:

    a) Magnetic North Pole _____

    b) Magnetic South Pole _____

3. Use 111 km (69 mi) per degree of latitude to find the linear distance in kilometers and miles between the latitude of the:

   a) Magnetic North Pole and true north _____

   b) Magnetic South Pole and true south _____

Topographic maps (or topo maps) are a set of maps covering the entire United States that show the horizontal position (latitude/longitude) of boundaries, land-use aspects, bodies of water, and economic and cultural features. Topographic maps also have a vertical component to show topography (configuration of the land surface), including slope and relief (the vertical difference in local landscape elevation). These fine details are shown through the use of elevation contour lines. A contour line connects all points at the same elevation. Elevations are shown above or below a vertical datum, or reference level, which usually is mean sea level. The contour interval is the vertical distance in elevation between two adjacent contour lines. Inside the front cover of this manual are the standard symbols commonly used on these topographic maps. These symbols and the colors used are standard on all USGS topographic maps: black for human constructions, blue for water features, brown for relief features and contours, pink for urbanized areas, and green for woodlands, orchards, brush, and the like. The margins of a topographic map contain a wealth of information about its concept and content. In the margins of topographic maps, you find the quadrangle name, names of adjoining quads, quad series and type, position in the latitude-longitude and other coordinate systems, title, legend, datum plane, symbols used for roads and trails, the dates and history of the survey of that particular quad, magnetic declination (alignment of magnetic north) and compass information, and more.

Magnetic declination is shown by a declination diagram or arrow in the margin of a topographic map, as shown in **Figure 3.4**. True north is indicated by the star, and the magnetic declination is shown by the arrow marked MN. You work further with magnetic declination in Lab Exercise 23 (topographic map reading). **Grid north (GN)** refers to the relation of the topo map to the Universal Transverse Mercator system (UTM). GN is another reference system based on rectangular map zones, so that as meridians taper toward the pole, the GN shows the relation of these converging lines and the rectangular grid.

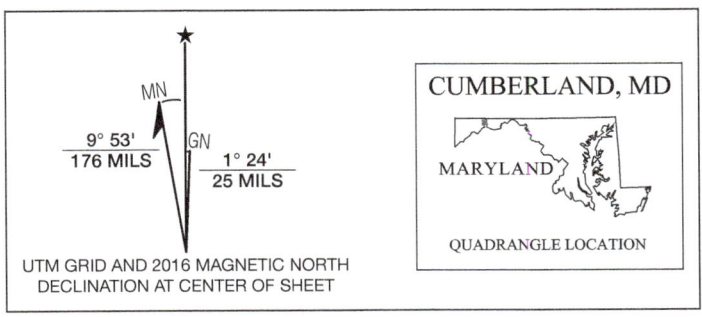

▲ **Figure 3.4** Declination arrow on topographic maps

4. According to Figure 3.4, what is the magnetic declination between:

   a) True north and magnetic north (MN)? _____

   b) True north and grid north (GN)? _____

5. Look for the declination arrow in the margin of the topographic map provided by your instructor. What is the magnetic declination between true north and magnetic north on the topographic map provided by your instructor? What is the magnetic declination between true north and magnetic north shown for this location in Figure 3.3? Any discrepancies you find relate to the migrating magnetic pole and the dates of the topo map and the isogonal map (2020). (Keep this topo map handy for the next section.)

SECTION 4

Compass Bearing

The compass bearing between two points on a map may be determined by using a protractor or a magnetic compass. To accomplish this, lightly draw a line through the specified points and extend the line to the edge of the map. To use a protractor, place the straight edge of the protractor along the margin of the map, with the origin of the protractor at the point where the line intersects the map margin. The extension of the drawn line indicates the compass bearing (**Figure 3.5a**). To use the compass, simply align the margin of the map with the 0°/360° and 180° points on the compass. The bearing can be read from the compass dial (**Figure 3.5b**). (Pay no attention to the compass needle.) The opposite direction from an azimuth is the back azimuth. To find the back azimuth of a bearing either add or subtract 180° from your bearing. If your bearing is less than 180°, then add 180°. If your bearing is more than 180°, then subtract 180° from your bearing. For example, if your bearing is 90°, due east, adding 180° to it results in a back azimuth of 270°, which is due west. If your bearing was 270°, you would subtract 180° from it to find a back azimuth of 90°.

▲ **Figure 3.5** Reading compass bearings on a map using a protractor (a); using a magnetic compass (b). (From Busch, Richard M., editor, *Laboratory Manual in Physical Geology*, 3rd ed., Macmillan Publishing Company © 1993.)

Using the local topographic map from the last section, complete the following.

1. What is the compass bearing between two points on the map indicated by your instructor? Give the reading as an azimuth. Repeat with the second and third sets of points as indicated by your instructor.

   a) From _____ to _____, azimuth: _____ back azimuth: _____

   b) From _____ to _____, azimuth: _____ back azimuth: _____

   c) From _____ to _____, azimuth: _____ back azimuth: _____

On the circle provided below, sketch your determination of the azimuths.

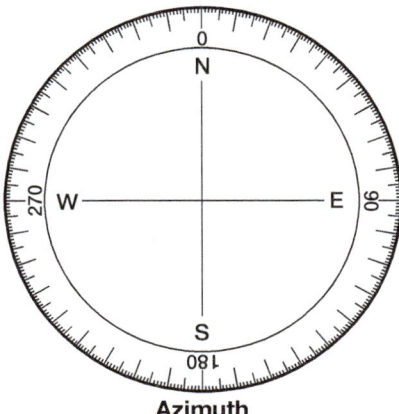

Azimuth

2. What is the relief (highest elevation minus lowest elevation in a local landscape) between two points in Question 1 above? (Show your work.)

   a) What is the relief between the two points in Question 1a above?

   b) What is the relief between the two points in Question 1b above?

   c) What is the relief between the two points in Question 1c above?

3. What is the highest elevation on the local topographic map? (Indicate a geographic feature that is at or adjacent to that elevation.)

4. What is the lowest elevation on the map? (Indicate a geographic feature that is at or adjacent to that elevation.)

5. What is the relief (highest elevation minus lowest elevation in a local landscape) of the area portrayed on this map? (Show your work.)

Mastering Geography

Looking for additional lab prep and review materials? Visit the Study Area in Mastering Geography to enhance your geographic literacy and spatial reasoning skills by accessing MapMaster™ interactive maps, videos, *Mobile Field Trips*, *Project Condor* Quadcopter Videos, *In the News* articles, flashcards, web links, self-study quizzes, and other resources. Contact your professor to find out how you can access Mastering Geography.

NOTES

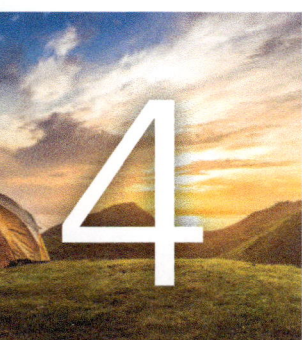

LAB EXERCISE

Map Projections, Map Reading, and Interpretation

A **map** is a generalized view of an area, usually some portion of Earth's surface, as seen from above and greatly reduced in size. The part of geography that embodies mapmaking is called **cartography**. Maps are critical tools with which geographers depict spatial information and analyze spatial relationships. We all use maps at some time to visualize our location and our relationship to other places, or maybe to plan a trip, or to coordinate commercial and economic activities.

In this lab exercise, we begin with an overview of various map projections that are used for a variety of purposes. An important decision in making maps is the proper scale to use. Maps, like architectural blueprints, are produced at a greatly reduced size compared to reality. Then, we specifically work with a U.S. mapping program—the township and range system. Lab Exercise 4 has four sections.

Key Terms and Concepts

cartography
conic
cylindrical
equal area (equivalent)
gnomonic projection

great circle
map
map projection
Mercator projection
planar

rhumb lines
scale
standard line
township and range
true shape (conformal)

KEY**LEARNING**concepts

After completion of this lab, you should be able to:

- **Define** the map projection concept and *identify* the four classes of map projections.
- **Recognize** distortions that are characteristic of selected map projections and *compare* distortions in order to *select* the appropriate map for a given use.
- **Plot** a great circle route from a gnomonic projection to a Mercator projection.
- **Compare** methods of expressing map scale and *calculate* map scales.
- **Explain** the township and range system employed in the United States.

Materials/Sources Needed

tracing (or wax) paper or notebook paper
world globe
scissors

color pencils
calculator
ruler

Lab Exercise and Activities

SECTION 1

Map Projections

We worked with Earth's coordinate grid system in Lab Exercise 2. Cartographers prepare large-scale flat maps—two-dimensional representations (scale models) of our three-dimensional Earth. This transformation of spherical Earth and its latitude–longitude coordinate grid system to a flat surface in some orderly and systematic realignment is called a **map projection**.

A globe is the only true representation of distance, direction, area, shape, and proximity, and preparation of a flat version requires decisions as to the type and amount of distortion that will be acceptable. To understand this problem, consider these important physical properties of a globe:

- Parallels are always parallel to each other, are always evenly spaced along meridians, and always decrease in length toward the poles.
- Meridians converge at both poles and are evenly spaced along any individual parallel.
- The distance between meridians decreases toward poles, with the spacing between meridians at the 60th parallel equal to one-half the equatorial spacing.
- Parallels and meridians always cross each other at right angles.

The problem with preserving these properties is that all globe qualities cannot be reproduced on a flat surface. The larger the area of Earth depicted on a map, the greater the distortion will be. Simply taking a globe apart and laying it flat on a table illustrates the problem cartographers face in constructing a flat map (**Figure 4.1**).

In Figure 4.1, you can see the empty spaces that open up between the sections, or gores, of the globe when it is flattened. Flat maps always possess some degree of distortion—much less for large-scale maps representing a few kilometers, and much more for small-scale maps covering individual countries, continents, or the entire world.

The best projection is always determined by its intended use. The major decisions in selecting a map projection involve the properties of **equal area** (equivalence) and **true shape** (conformality). If a cartographer selects equal area as the desired trait, as for a map showing the distribution of world climates, then shape must be sacrificed by *stretching* and *shearing* (having parallels and meridians cross at other than right angles).

If, on the other hand, true shape is desired, as for a map used for navigational purposes, then equal area must be sacrificed, and the scale will actually change from one region of the map to another. Two additional properties related to map projections are true direction and true distance.

The fold-out flap of the back cover shows four classes of map projections and geometric-surface perspectives from which three classes of maps—**cylindrical**, **planar** (or *azimuthal*), and **conic**—are generated. Another class of projections that cannot be derived from this physical-perspective approach is the nonperspective *oval*-shaped. Still others are derived from purely mathematical calculations.

With all projections, the contact line (or point) between the globe and the projection surface—called a **standard line**—is *the only place where all globe properties are preserved*. A *standard parallel* or *standard meridian* is true to scale along its entire length, without any distortion. Areas away from this critical tangent line or tangent point become increasingly distorted. Consequently, this area of optimum spatial properties should be centered on the region of immediate interest so that greatest accuracy is preserved there.

Animation
Map Projections

https://goo.gl/0iVdDM

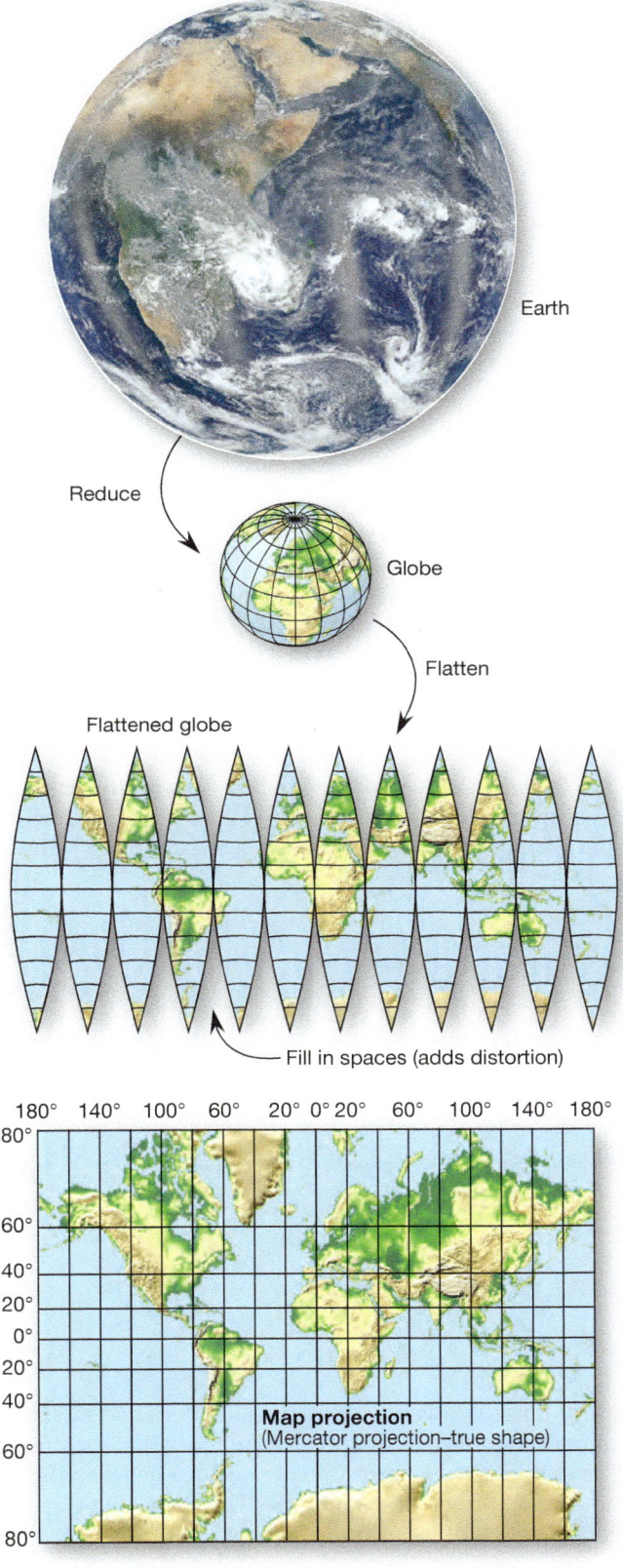

Earth

Reduce

Globe

Flatten

Flattened globe

Fill in spaces (adds distortion)

180° 140° 100° 60° 20° 0° 20° 60° 100° 140° 180°

80°
60°
40°
20°
0°
20°
40°
60°
80°

Map projection
(Mercator projection–true shape)

▲ **Figure 4.1** Conversion of a globe to a flat map requires decisions about the desired properties to preserve, how much distortion is acceptable, and which projection best serves the purpose of the map. [NASA/NOAA/GSFC/Suomi NPP/VIIRS/Norman Kuring]

SECTION 2

Cylindrical, Planar, and Conic Projections

This step may be done by working in groups. Your instructor will direct you.

1. Using a globe and tracing (or wax) paper—notebook paper works with an illuminated globe—trace outlines of North America, South America, and Greenland. You can do this quickly; a rough outline showing size and shape is sufficient.

    Using a globe and your tracings, compare the relative sizes of North America, South America, and Greenland. Which is larger and by approximately how many times? Describe your observations.

2. Examples of two cylindrical projections are presented in **Figure 4.2**, the **Mercator projection** (conformal, true-shape), and **Figure 4.3**, the *Lambert projection* (equivalent, equal-area). These two map projections—the Mercator and the Lambert—are simply being used to examine *relative* size and shape in each portrayal of Earth.

    Trace around the outlines of the same continents and island as indicated above—North America, South America, and Greenland. You might want to mark and color code the tracings according to their source (Mercator or Lambert) for future reference. Use your tracings to answer the following questions. Keep in mind that these two sets of map outlines compared to the globe you used are not at the same scale: the distance measured along their respective equators will not be equal.

    a) Using the outlines from the Mercator and Lambert map projections on the next pages, make the same comparisons among North America, South America, and Greenland. Which is relatively larger and by approximately how many times? Again, describe your observations.

    Mercator

    Lambert

    b) Now compare the relative shapes of Greenland, North America, and South America from the globe (shows true shape) with those from the Mercator and Lambert projections. Is there distortion in terms of shape? If so, briefly explain how they are distorted.

    Mercator

    Lambert

    Figure 4.4a and **Figure 4.4b** once again show the Mercator and Lambert cylindrical projections. Diagrams to the right of each projection help you to identify the distortion in each. The circles along the equator—which is the standard line, or line of tangency between the globe and the cylinder and is without distortion—are the "reference standard." The shape and size of the other "circles" is proportionate to the distortion at various latitudes on the maps.

10° graticule (the latitude-longitude grid of parallels and meridians); central meridian is 90°W

▲ **Figure 4.2** Mercator projection—conformal, true-shape map

▲ **Figure 4.3** Lambert projection—equivalent, equal-area map

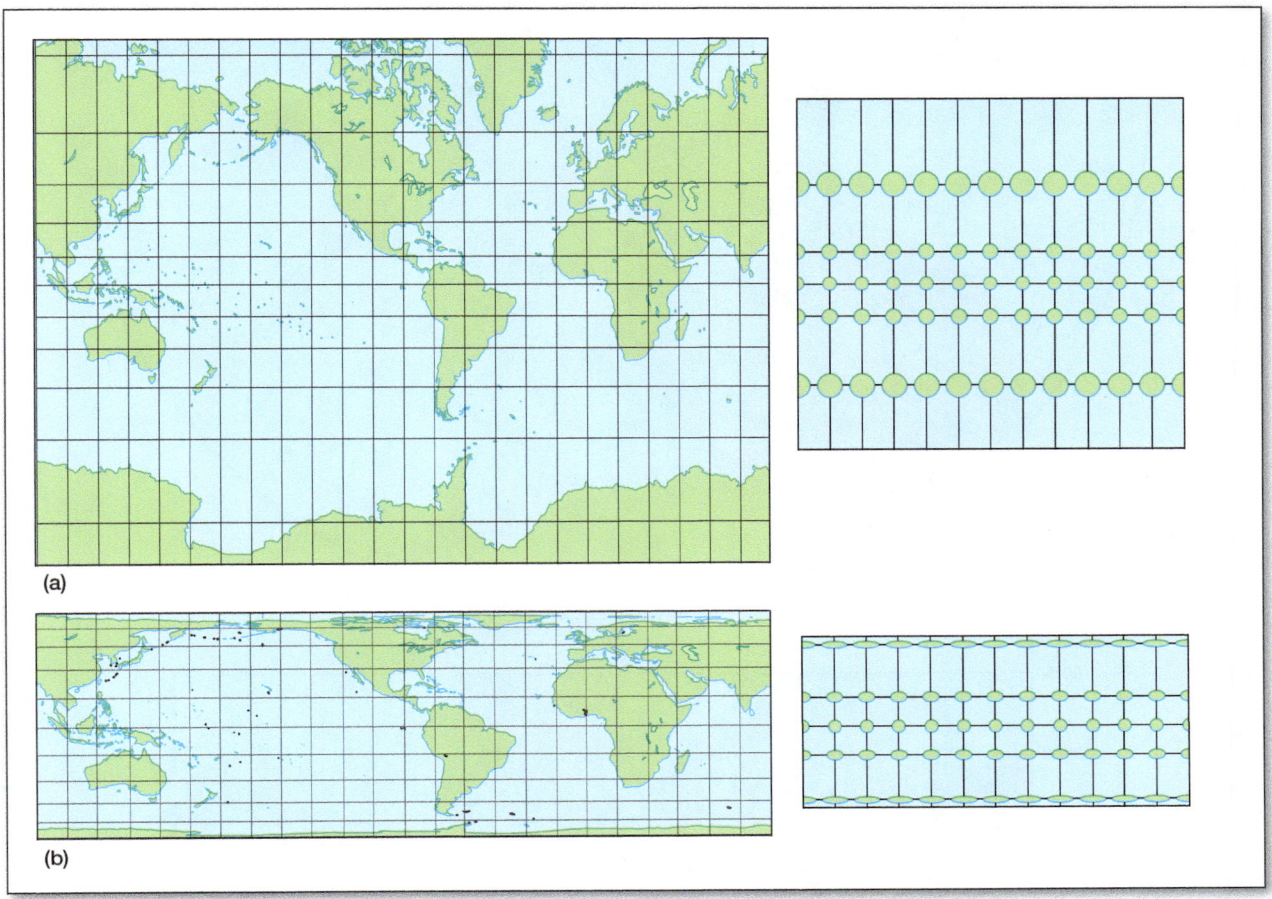

▲ **Figure 4.4** (a) Mercator cylindrical true shape (conformal) projection, 30° graticule; standard parallel 0° central meridian 90°W.
(b) Lambert cylindrical equal area (equivalent) projection, 30° graticule; standard parallel 0° central meridian 90°W.

3. What happens to the circles in the Mercator projection (and, therefore, to the landmasses) as latitude increases? What causes this distortion of the circles away from the equator? (*Hint:* Compare parallels and meridians with those on the globe.)

4. What happens to the circles in the Lambert projection (and, therefore, to the landmasses) as latitude increases? What causes this distortion of the circles away from the equator? (Once again, compare parallels and meridians.)

5. Cylindrical projections such as Mercator and Lambert would best be used in mapping what areas of the globe? Why?

▲ **Figure 4.5** (a) Gnomonic/planar projection; (b) Mercator/cylindrical projection.

6. **Figure 4.5a** shows Earth on a *planar* projection, in which the globe is projected onto a plane. This is a **gnomonic projection** and is generated by projecting a light source at the center of a globe onto a plane that is tangent to (touching) the globe's surface. The resulting increasingly severe distortion as distance increases from the standard point prevents showing a full hemisphere on one projection.

    a) Where is the projection surface tangent to (touching) the globe? (See end flap of this lab manual.)

    b) What kind of distortion, if any, occurs on a planar projection, and where?

c) Can you show the entire Earth on a single gnomonic projection? If not, why not?

d) Which areas of the globe would likely be mapped on a planar projection, and why?

The Mercator projection is useful in navigation and has become the standard for nautical charts since Gerardus Mercator, a Flemish cartographer, devised it for navigational purposes in 1569. The advantage of the Mercator projection is that lines of constant compass direction, called **rhumb lines**, are straight and thus facilitate the task of plotting compass directions between two points. However, these lines of constant direction, or bearing, are not the shortest distance between two points.

From the *planar* class of projections, a gnomonic projection possesses a valuable feature: All straight lines are **great circles**—the shortest distances between two points on Earth's surface. A gnomonic projection is used to determine the coordinates of a great-circle (shortest) route; these coordinates are then transferred to a Mercator (true-direction) projection for determination of precise compass headings—the route along which the pilot or captain steers the airplane or ship.

7. Use the two maps in Figure 4.5 to plot a great circle route between San Francisco (west coast of the United States, where you can see small details of San Francisco Bay) and London (southern England at the 0° prime meridian). The straight line on the gnomonic projection will show the shortest route between the two cities. Transfer the coordinates of this route over to the Mercator map and connect with a line plot to show the route's track. Note the route arching over southern Greenland on the Mercator.

8. Assume you board a plane in San Francisco for a flight to London. Briefly describe your great-circle flight route between the two cities (plotted on the map in Figure 4.5b). What are some of the features over which you fly? Describe the landscape and water below.

SECTION 3

Map Scale

The ratio of the measured distance on a map to the actual distance on the ground is called **scale**; it relates a unit on the map to a similar unit on the ground. A 1:1 scale would mean that a centimeter on the map represents a centimeter on the ground, or an inch on the map represents an inch on the ground. A more appropriate scale for a local map is 1:24,000, in which 1 unit on the map represents 24,000 identical units on the ground.

Map scales may be presented in several ways. A *written* (or verbal) *scale* simply states the ratio using words—for example, "one centimeter to one kilometer" or "one inch to one mile." A *representative fraction* (RF, or fractional scale) can be expressed with either a ":" or a "/", as in 1:125,000 or 1/125,000. No actual units of measurement are mentioned because any unit is applicable as long as both parts of the fraction are in the same unit: 1 cm to 125,000 cm, 1 in. to 125,000 in., or even 1 arm length to 125,000 arm lengths, and so on. A *graphic* (bar) *scale* is a line or a bar divided into segments illustrating the correct map-to-ground ratio (**Figure 4.6**).

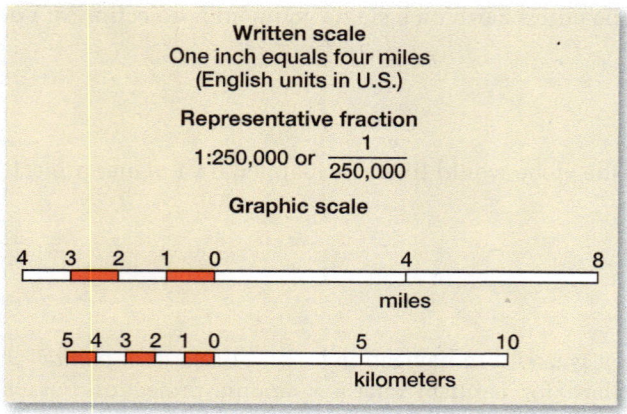

▲ **Figure 4.6** Three expressions of map scale: written scale, representative fraction, and graphic (bar) scale

Scales are called "small," "medium," and "large," depending on the ratio described. Thus, a scale of 1:24,000 is a large scale, whereas a scale of 1:50,000,000 is a small scale. The greater the size of the denominator in a fractional scale (or number on the right in a ratio scale), the smaller the scale and the more abstract the map must be in relation to what is being mapped.

Examples of selected representative fractions and written scales are listed in **Table 4.1** for small-, medium-, and large-scale maps.

TABLE 4.1 Sample representative fractions and written scales for small-, medium-, and large-scale maps			
System	Scale Size	Representative Fraction	Written Scale
English	Small	1:3,168,000	1 in. = 50 mi
		1:2,500,000	1 in. = 40 mi
		1:1,000,000	1 in. = 16 mi
		1:500,000	1 in. = 8 mi
		1:250,000	1 in. = 4 mi
	Medium	1:125,000	1 in. = 2 mi
		1:63,360 (or 1:62,500)	1 in. = 1 mi
		1:31,680	1 in. = 0.5 mi
		1:30,000	1 in. = 2500 ft
	Large	1:24,000	1 in. = 2000 ft
System		Representative Fraction	Written Scale
Metric		1:1,000,000	1 cm = 10.0 km
		1:50,000	1 cm = 0.50 km
		1:25,000	1 cm = 0.25 km
		1:20,000	1 cm = 0.20 km

Assume:

If a world globe is 61 cm (24 in.) in diameter and we know Earth has an equatorial diameter of 12,756 km (7926 mi), then the scale of the globe is the ratio of 61 cm to 12,756 km. Therefore, in order to determine the globe's representative fraction:

$$\mathbf{RF} = \frac{\text{diameter of globe}}{\text{diameter of Earth}} = \frac{61 \text{ cm}}{12,756 \text{ km}} = \frac{61 \text{ cm}}{1,275,600,000 \text{ cm}} = \frac{1 \text{ cm}}{20,911,000 \text{ cm}}$$

Thus, the representative fraction for the 61 cm globe is expressed in centimeters as 1:20,900,000 (rounded off). This representative fraction can now be expressed in *any* unit of measure, metric or English, as long as both numbers are in the same units.

Complete:

1. Several companies manufacture large world globes for display in museums, corporate lobbies, and science exhibit halls. If such a globe featured a *diameter* of 4 m (13 ft), what is the scale of this globe expressed as a representative fraction? (Show your work.) (*Hint:* Earth's equatorial diameter = 12,756 km [7926 miles].)

2. Given the large globe in Question 1 and your calculation of its representative fraction, convert the RF to a *written scale* for this globe (1 cm on the globe = ? km on Earth). (Show your work and remember that there are 100,000 cm in 1 km.)

3. Use the following topographic maps and a ruler (inches or cm) to measure the map distance between the points indicated. Noting the RF scales posted on each map, calculate the approximate ground distance (miles or kilometers). Keep in mind that there are 63,360 inches in a mile: 12 in./ft × 5280 ft/ mile. Compare results with the graphic scale. (Show your work.)

    a) **Figure 20.6 Kīlauea Crater, HI, Topographic Map 1:** from the 3700-ft. contour label south of the Uēkahuna Bluff label to the Byron Ledge label on the east side of the crater.

    b) **Figure 25.6 Mt. Rainier National Park Topographic Map:** from Columbia Crest to McClure Rock.

    c) **Figure 26.6 Death Valley Topographic Map:** from the 1000′ contour line in Trail Canyon on the west side of Death Valley to the 200′ contour line on the east side of Death Valley.

4. Find three different maps or globes as follows. Briefly describe the content of each and record the scale as a representative fraction or other scale noted on each map or globe.

    a) Classroom globe _____

    b) Map from your atlas _____

    c) Topo map at back of this manual _____

5. Give an example of a specific scale for each of the following:

    a) Small scale: _____

    b) Medium scale: _____

    c) Large scale: _____

6. What is an appropriate scale to use in preparing a world map for a classroom (pull-down "wall" map)? Why?

7. If you wanted a map to help you plan a local trip within your state or province, what scale would be best?

8. If you were helping to plan the painting of a world map on a local school playground that is approximately 10 meters along the equator, what scale would be best?

RF: _____

Written scale: _____

9. Using the metric-to-English conversion chart on the inside of the fold-out back flap of this manual, what would be the approximate length of the equator, in feet?

SECTION 4

Township and Range Survey System

The Public Lands Survey System (1785) delineated locations in the United States. After modifications and adjustments, the final plan called for the establishment of the **township and range** grid system, based on an initial survey point. For those portions of the United States surveyed under this system, the *initial points* are presented in **Figure 4.7**. Places are designated as north or south of the base line and east or west of the principal meridian. The names of the base lines and meridians are the same for the same initial point. For example, Washington state uses the Wilamette Base Line and Meridian. For example, Township 3 N means 3 "rows" north of the base line, and Range 4 W means 4 "columns" west

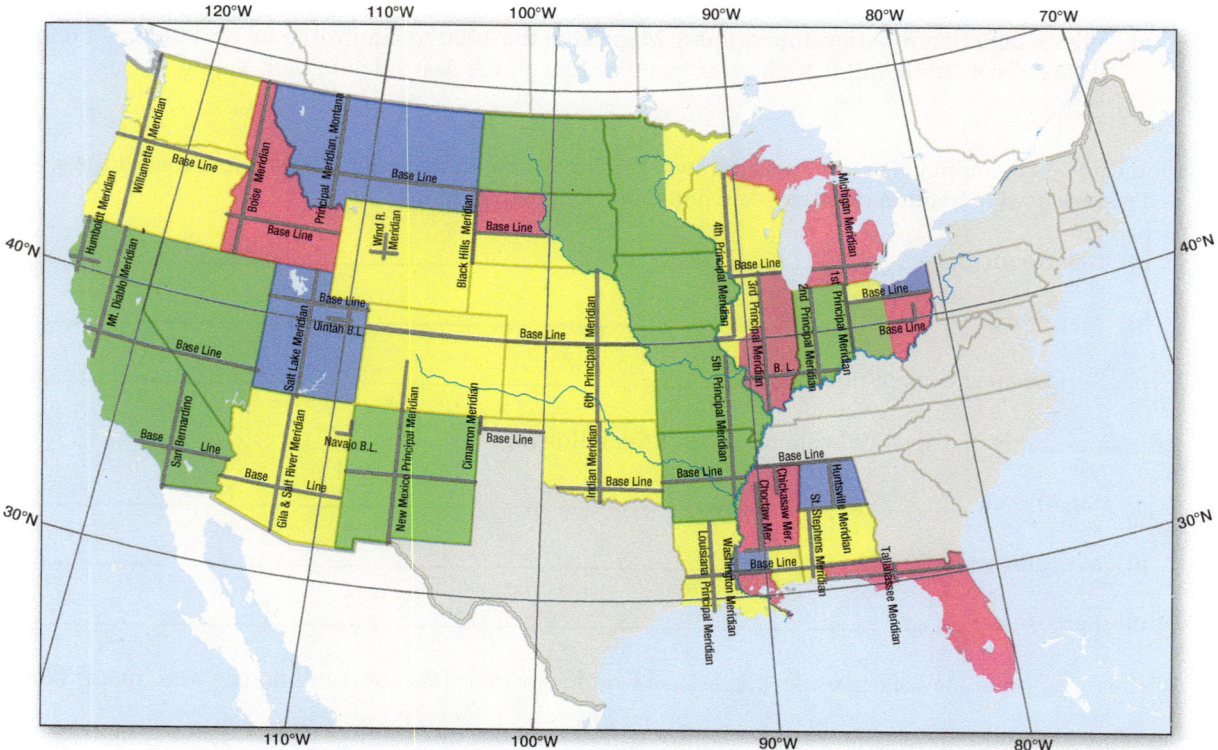

▲ **Figure 4.7** Principal meridians and base lines used for the township and range public-land surveys. Shaded states were not surveyed with the PLSS. (Adapted from the Bureau of Land Management, *Manual of Surveying Instructions*, Washington, DC, 1947, p. 168.)

ROUTLEDGE

STUDY GUIDE

This
Study Guide
for all members
of a study group
and for the
facilitator.

Third Edition

WHAT
GREAT TEACHERS
DO DIFFERENTLY

Nineteen Things That Matter Most

TODD WHITAKER

Executive Editor, Geosciences Courseware: Christian Botting
Courseware Director, Portfolio Management: Beth Wilbur
Managing Producer, Science: Mike Early
Content Producer: Anton Yakovlev
Production Management: Jeanine Furino, Cenveo Publisher Services
Editorial Assistant, Geosciences Courseware: Emily Bornhop
Executive Field Marketing Manager: Mary Salzman
Illustrators: Lachina and International Mapping Associates
Senior Content Producer: Tim Hainley
Cover Photo: *Island in Lake Baikal*, by Chalermkiat Deedokmai/Getty Images

PEARSON, ALWAYS LEARNING and MasteringGeography are exclusive trademarks in the U.S. and/or other countries owned by Pearson Education, Inc. or its affiliates.
Google Earth is a trademark of Google. ArcGIS is a trademark of ESRI.

Photo Credits: Opener Images: Chapters 1-11, 30: FotoVoyager/E+/Getty Images. Chapters 12-18: Michael & Jennifer Lewis/Getty Images. Chapters 19-27: David Noton Photography/Alamy. Chapters 28-29: Paul Baggaley/Getty Images.

Many of the designations used by manufacturers and sellers to distinguish their products are claimed as trademarks.
Where those designations appear in this book, and the publisher was aware of a trademark claim, the designations have been printed in initial caps or all caps.

Library of Congress Cataloging-in-Publication Data

Names: Thomsen, Charles E., author. | Christopherson, Robert W., author.
Title: Applied physical geography : geosystems in the laboratory / Charles E.
    Thomsen, American River College, Robert W. Christopherson, American River
    College--Emeritus.
Other titles: Geosystems
Description: Tenth edition. | Boston : Pearson, [2017] | Title of previous
    edition: Geosystems : an introduction to physical geography. | Previous
    edition entered under: Christopherson, Robert W.
Identifiers: LCCN 2017024219| ISBN 9780134686363 (alk. paper) | ISBN
    0134686365 (alk. paper)
Subjects: LCSH: Physical geography.
Classification: LCC GB54.5 .C48 2017 | DDC 910/.02--dc23
LC record available at https://lccn.loc.gov/2017024219

ISBN-13: 978-0-134-68636-3
ISBN-10: 0-134-68636-5

Second, you will gather data for up to 30 days (or less, as assigned by your instructor) during the term, not necessarily consecutive days, and record them on a weather calendar using official international symbols and designations. If your lab meets 15 or 16 times during the term and the department has weather instruments, then half your days of observation are easily completed on class days. The other days can be met by outside sources or visits to the department.

The National Weather Service and the Canadian Meteorological Center use a standardized system of recording and reporting weather data around a weather station symbol developed by the World Meteorological Organization (WMO). An example of this method is shown below.

WEATHER STATION SYMBOL

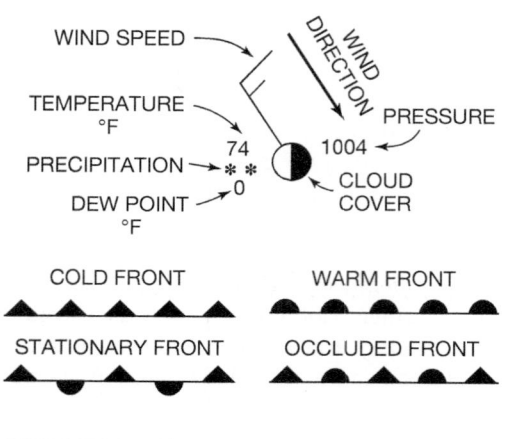

COLD FRONT WARM FRONT

STATIONARY FRONT OCCLUDED FRONT

PRECIPITATION TYPE

, DRIZZLE
• RAIN
✳ SNOW
▽ SHOWERS
T̄ THUNDERSTORMS
≡ FOG
∞ DRY HAZE
⌒⌣ FREEZING RAIN
△ HAIL
△ SLEET

WIND SPEED
(kph/mph)

CALM	◎	
5–13		3–8
14–23		9–14
24–32		15–20
33–40		21–25
89–97		55–60
116–124		72–77

CLOUD COVER

○	NO CLOUDS	⊕	SIX-TENTHS
◐	ONE-TENTH OR LESS	◕	SEVEN-TENTHS TO EIGHT-TENTHS
◔	TWO-TENTHS TO THREE-TENTHS		NINE-TENTHS OR OVERCAST WITH OPENINGS
◔	FOUR-TENTHS	◖◗	
◑	FIVE-TENTHS	●	COMPLETELY OVERCAST
		⊗	SKY OBSCURED

Attached is your "Weather Calendar," with spaces for up to 30 days of weather observations. These days need not be consecutive and may be spread out over the entire term. The completed calendar is due at the end of the term. Your instructor will give you more details. The circle in the middle of each square is the weather station symbol, and above and below are spaces to list the date and the place of observation. (If you are using a broadcast source, list the place the weather instruments are located that made the observation.)

We can refer to the weather station symbol as if it were a clock. Temperature is recorded in the 10:00 position, barometric pressure at 2:00, weather type at 9:00, and dew-point temperature (if available) at 8:00; wind speed and direction vary in location, depending on the wind itself. If you are observing clouds, high clouds are symbolized above the station symbol, and middle and low clouds below it.

The state of the sky (cloud cover) is the area inside the circle. In the example to the left, temperature is 74°F (23°C), pressure is 1004.0 mb, the sky is 50% (five-tenths) overcast, with slight continuous fall of snowflakes at time of observation. Winds are from the northwest (line drawn on the side of the station where the winds originate) at 15–20 mph (24–32 kmph), as noted by the "flags" on the line.

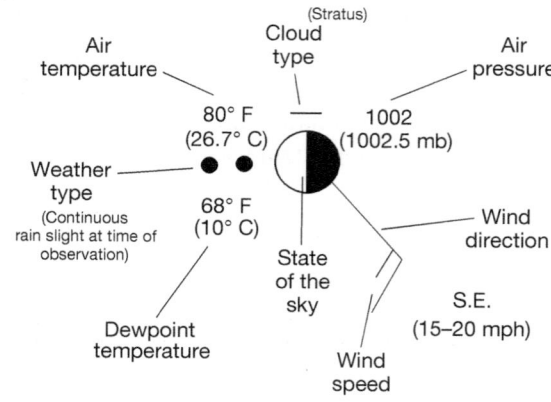

SECTION 3

Water Budget Calculations for Sacramento, California

For comparison, let's work with a city that experiences seasonal deficits in its annual water balance. Sacramento, California (38.58°N, 121.35°W, at 11 m elevation), has a Mediterranean dry, warm summer climate. The data for Sacramento are set in **Table 16.2**. Please assume the same soil-moisture storage capacity of 100 mm (4.0 in.) for Sacramento as for Kingsport, typical of shallow-rooted plants. The months of net demand for moisture are satisfied through *soil-moisture utilization*, as long as the soil moisture is available. Sacramento does experience a wilting point each year.

TABLE 16.2 Water balance for Sacramento, California

	Jan	Feb	Mar	Apr	May	Jun	Jul	Aug	Sep	Oct	Nov	Dec	Total
PRECIP	107	74	56	36	10	3	3	3	8	23	58	76	455
POTET	15	23	40	60	85	115	140	127	99	66	30	15	815
PRECIP − POTET													—
STRGE	100					0						89	—
ΔSTRGE													
ACTET													
DEFIC													
SURPL													

(All quantities in millimeters)

Questions and analysis about the water balance of Sacramento, California:

1. Complete the bookkeeping procedure for Sacramento using the monthly and annual average values given in Table 16.2. Calculate PRECIP − POTET, ΔSTRGE, ACTET, DEFIC, and SURPL amounts for each month, as you did for Kingsport.

2. For Sacramento, how many months does POTET exceed PRECIP?

Mastering Geography

Looking for additional lab prep and review materials? Visit the Study Area in Mastering Geography to enhance your geographic literacy and spatial reasoning skills by accessing MapMaster™ interactive maps, videos, *Mobile Field Trips*, *Project Condor* Quadcopter Videos, *In the News* articles, flashcards, web links, self-study quizzes, and other resources. Contact your professor to find out how you can access Mastering Geography.

NOTES

Video
Exercise 17
Pre-Lab Video

http://goo.gl/9zDN70
Scan to view the
Pre-Lab video

LAB EXERCISE

Global Climate Systems

Earth experiences an almost infinite variety of *weather*—conditions of the atmosphere at any given time and place. But if we consider the weather over many years, including its variability and extremes, a pattern emerges that constitutes **climate**. Think of climate patterns as dynamic rather than static. Climate is more than a consideration of simple averages of temperature and precipitation.

Climatology, the study of climate, is the analysis of long-term weather patterns, including extreme weather events, over time and space to find areas of similar weather statistics, identified as **climatic regions**. Observed patterns grouped into regions are at the core of climate classification.

In this lab exercise, we examine various patterns of temperature and precipitation that operate as a basis for climate classification. We work with a model of global climates, the classification and plotting of actual climate data, and analysis of temperature and precipitation patterns. Lab Exercise 17 features five sections (one optional).

Key Terms and Concepts

classification
climate
climatic regions

climatology
climographs

empirical classification
genetic classification

KEY**LEARNING**concepts

After completion of this lab, you should be able to:

- *Identify* patterns of temperature, precipitation, and other weather elements that contribute to the climate of an area.
- *Analyze* key climate data to *determine* the empirical (Köppen) classification and genetic (causative) factors.
- *Plot* data (°C and cm) on climographs for analysis.

Materials/Sources Needed

colored pencils
calculator
world atlas

physical geography textbook
biomes chart in Exercise 29

Lab Exercise and Activities

SECTION 1

Climatology and Climate Classification

Climatology, the study of climate, involves analysis of the patterns in time or space created by various physical factors in the environment. One type of climatic analysis involves discerning areas of similar weather statistics and grouping them into climatic regions that contain characteristic weather patterns.

Climate classifications are an effort to formalize these patterns and determine their implications for humans. Simply comparing the two principal climatic components—temperature and precipitation—reveals important relationships that are indicative of the distribution of climate types (**Figure 17.1**).

▲ **Figure 17.1** Temperature and precipitation schematic showing climatic relationships

1. Mark your present location with an **X** on Figure 17.1. Place the name of this location next to the **X**. Next, mark the location (approximately) that is most characteristic of your place of birth with an **O** and record the name of that place next to the **O**.

Classification is the process of grouping data or phenomena in related classes. A classification based on causative factors—for example, the interaction of air masses—is called a **genetic classification**. An **empirical classification** is based on statistics or other data used to determine general categories. Climate classifications based on temperature and precipitation data are empirical classifications. In *Applied Physical Geography,* we use Köppen's empirical system to classify climatic regions, as well as a genetic analysis to help you appreciate the relationships between a region's location and its climate.

The principal elements of climate are insolation, temperature, pressure, air masses, and precipitation. We review them briefly here. Insolation is the energy input for the climate system, but it varies widely over Earth's surface by latitude (see Figure 6.2). The pattern of world temperatures and their annual ranges is shown in Figures 9.2a and 9.2b. Temperature variations result from a coupling of dynamic forces in the atmosphere to Earth's pattern of atmospheric pressure and resulting global wind systems (see Figures 11.6 and 11.7). Important, too, are the location and physical characteristics of air masses, those vast bodies of homogeneous air that form over oceanic and continental source regions.

Moisture is the remaining input to climate. The hydrologic cycle transfers moisture, with its tremendous latent heat energy, through Earth's climate system. Moisture patterns are important, for they are key climate control factors. Average temperatures and daylength help us approximate POTET (potential evapotranspiration), a measure of natural moisture demand. Most of Earth's desert regions, areas of permanent water deficit, are in lands dominated by subtropical high-pressure cells, with bordering lands grading to grasslands and to forests as precipitation increases. The most consistently wet climates on Earth straddle the equator in the Amazon region of South America, the Congo region of Africa, and Indonesia and Southeast Asia, all of which are influenced by equatorial low pressure and the intertropical convergence zone, or ITCZ. Simply relating the two principal climatic components—temperature and precipitation—reveals general climate types.

The basis of any empirical classification system is the choice of criteria used to draw lines between categories. The Köppen classification system, widely used for its ease of comprehension, was designed by Wladimir Köppen (1846–1940), a German climatologist and botanist. Köppen chose mean monthly temperatures, mean monthly precipitation, and total annual precipitation to devise his spatial categories and boundaries. But we must remember that boundaries really are gray areas; they are transition zones of gradual change. For our purposes of general understanding, a modified Köppen-Geiger classification is useful.

The Köppen system uses capital letters (A, B, C, D, E, plus H) to designate climatic categories from the equator to the poles. Five of the climate classifications are based on thermal criteria and one on moisture conditions as well.

A—Tropical (equatorial regions)

C—Mesothermal (Mediterranean, humid subtropical, marine west coast regions)

D—Microthermal (humid continental, subarctic regions)

E—Polar (polar regions)

H—Highland (compared to lowlands at the same latitude, highlands have lower temperatures—recall the normal lapse rate—and more efficient precipitation due to lower POTET demand, justifying separation of highlands from surrounding climates)

B—Dry, the deserts and steppes (only category based on moisture as well)

In addition to these primary types noted with capital letters, lowercase letters are used to signify certain temperature and moisture characteristics. Inside the back cover of this laboratory manual is a world map illustrating climates classified according to the Köppen system, with the addition of the main climatic influences. Table 17.1 presents a climate classification key and summary that specifies both *thermal* (A, C, D, and E) and *moisture* (B) classifications. Use this table for classifying each of the five stations presented in Lab Exercise Section 2 and two stations in Section 4. Specific formulas to calculate the *B* dry climates follow in this lab exercise in Section 4.

For example, in a *tropical rain forest Af* climate, the *A* tells us that the average coolest month is above 18°C (64.4°F, average for the month), and the *f* indicates that the weather is constantly wet (German *feucht*, for "moist"), with the driest month receiving at least 6 cm (2.4 in.) of precipitation.

As you can see in the map of *A* climates, the *tropical rain forest Af* climate straddles the equator.

For another example, in a *Dfa* climate, the *D* means that the average warmest month is above 10°C (50°F), with at least one month falling below 0°C (32°F); the *f* says that at least 3 cm (1.2 in.) of precipitation falls during every month; and the *a* indicates a warmest summer month averaging above 22°C (71.6°F). Thus, a *Dfa* climate is a humid-continental, hot-summer climate in the microthermal D category.

Try not to get lost in the alphabet soup of this system, for it is meant to help you understand complex climates through a simplified set of symbols. To simplify the detailed criteria, a box is placed at the end of each climate category that presents Köppen's guidelines. To help keep it straight, always say the climate's descriptive name followed by its symbol: for example, *Mediterranean dry-summer Csa, tundra ET*, or *cold midlatitude steppe BSk*.

Determining primary climate classification

E — An **E climate** if warmest month is less than 10° C (50° F). If not an *E* then move on to *B* climates.

B — If precipitation supply is less than average potential evapotranspiration demand climate may be a **B climate**—go to Lab **SECTION 4** to consult formulas. If climate is more moist than *B* criteria then move on to *A* climates.

A — An **A climate** if the average coolest month is warmer than 18° C (64.4° F). If not an *A* then move on to *C* climates.

C — A **C climate** if average coldest month is below 18° C (64.4° F) but warmer than 0° C (32° F). If not an *E, B, A,* or *C* then climate is a *D* classification.

D — A **D climate** if average coldest month is below 0° C (32° F) and warmest month is above 10° C (50° F).

E Polar Climates

Warmest month below 10° C (50° F); always cold; ice climates.

EF - Ice cap:
Warmest month between 0° C (32° F); PRECIP exceeds very small POTET demand; the polar regions.

ET - Tundra:
Warmest month 0–10° C (32–50° F); PRECIP exceeds small POTET demand; snow cover 8–10 months.

EM - Polar marine:
All months above −7° C (20° F), warmest month above 0° C; annual temperature range <30° F (17° C).

B Dry Arid and Semiarid Climates

POTET exceeds PRECIP in all B climates. Subdivisions based on PRECIP timing and amount and mean annual temperature. Boundaries determined by formulas and graphs.

Earth's arid climates.	Earth's semiarid climates.
BWh - Hot low-latitude desert	BSh - Hot low-latitude steppe
BWk - Cold midlatitude desert	BSk - Cold midlatitude steppe
BW = PRECIP less than 1/2 POTET.	BS = PRECIP MORE THAN 1/2 POTET but not equal to it.
h = Mean annual temperature >18° C (64.4° F).	h = Mean annual temperature >18° C.
k = Mean annual temperature < 18° C.	k = Mean annual temperature < 18° C.

A Tropical Climates

Consistently warm with all months averaging above 18° C (64.4° F); annual PRECIP exceeds POTET.

Am - Tropical monsoon:
m = Short dry season, annual PRECIP is greater than 25 times (10 – PRECIP of driest month in cm), ITCZ 6–12 months dominant.

Af - Tropical rain forest:
f = All months receive PRECIP in excess of 6 cm (2.4 in.).

Aw - Tropical savanna:
w = Summer wet season, winter dry season, dryest month receives less than 6 cm PRECIP.

C Mesothermal Climates

Warmest month above 10° C (50° F); coldest month above 0° C (32° F) but below 18° C (64.4° F); seasonal climates.

Cfb, Cfc - Marine west coast: Mild-to-cool summer.
f = Receives year-round PRECIP
b = Mild summer; not (a), at least 4 months above 10° C.
c = Not (b), coldest month above −38° C.

Cfa, Cwa - Humid subtropical:
a = Hot summer: warmest month above 22° C (71.6° F).
w = Winter drought, wettest summer month 10 times more PRECIP than driest winter month.
f = Not (w), year-round PRECIP.

Csa, Csb - Mediterranean summer dry:
s = Summer drought, driest summer month receives less than 6 cm PRECIP and also less than 1/3 of PRECIP of wettest winter month.
a = Hot summer with warmest month above 22° C (71.6° F).
b = Mild summer; not (a), at least 4 months above 10°C.

D Microthermal Climates

Warmest month above 10° C (50° F); coldest month below 0° C (32° F); cool temperate-to-cold conditions; snow climates. In Southern Hemisphere, only in highland climates.

Dfb, Dwb - Humid continental:
b = Mild summer; not (a), at least 4 months above 10° C.
w = Winter drought, wettest summer month 10 times more PRECIP than driest winter month.
f = Not (w), year-round PRECIP.

Dfa, Dwa - Humid continental:
a = Hot summer; warmest month above 22° C (71.6° F).
w = Winter drought, wettest summer month 10 times more PRECIP than driest winter month.
f = Not (w), year-round PRECIP.

Dfc, Dwc, Dwd - Subarctic: Cool summers, cold winters.
w = Winter drought, wettest summer month 10 times more PRECIP than driest winter month.
f = Not (w), year-round PRECIP.
c = Not (b), coldest month above −38° C.
d = Not (c), coldest month below −38° C (36.4° F), in Siberia only.

TABLE 17.1 Climate classification key

SECTION 2

Climographs for Five Stations

Temperature patterns are influenced by several principal controls: latitudinal position, altitude, cloud cover, land–water heating differences, ocean currents and sea-surface temperatures, and general surface conditions. Review your work in Lab Exercise 8 with stations in Bolivia and Canada where you identified the effects of these controls.

This exercise features **climographs**, which show temperatures as a line graph and precipitation as a bar graph. Use the five climographs (**Figures 17.2a** through **17.2e**) and sets of data that follow to plot mean monthly temperature (*solid-line graph*), mean monthly precipitation (*a bar graph*), and potential evapotranspiration (*a dotted line*) data for the following stations. Use Table 17.1 to determine each station's climatic classification and then determine the city that identifies each station. Using your atlas and the map inside the back cover of this laboratory manual, find and record the names of the five other cities listed below. Use the world climate map inside the back cover of this laboratory manual to identify the climatic influences: high or low air pressure; types of air masses; degree of continentality; and warm and cool currents. Analyze the distribution of temperature and precipitation during the year and describe them in general terms: Is the temperature range large or small? Is it consistently warm or consistently cool? Is it dry year round, or is it wet year round? Which months are rainy, and which months are dry? Use the terrestrial biomes map on the back cover of this laboratory manual to identify the representative biome for each region.

Selected Global Climate Stations

Section 2—Tropical, mesothermal, and microthermal climographs, six stations.

> **Figure 17.2** Salvador (Bahia), Brazil: pop. 2,676,000; elev. 9 m (30 ft), completed as an example.

> **Figures 17.2a through 17.2e**—five stations to graph and identify (listed in alphabetical order)

>> Edinburgh, Scotland: pop. 478,000; elev. 134 m (439 ft), 56°N 3.1°W

>> Montreal, Québec, Canada: pop. 1,621,000; elev. 57 m (187 ft), 45.5°N 73.5°W

>> New Orleans, Louisiana: pop. 337,000; elev. 3 m (9 ft), 30°N 90°W

>> Oymyakon, Siberia, Russia: pop. 521; elev. 726 m (2382 ft), 63°N 143°E

>> Sacramento, California: pop. 486,000; elev. 11 m (36 ft), 38.5°N 121.3°W

Section 3—Climate analysis

Section 4—Desert climate climographs, two stations

> **Figures 17.4a and 17.4b**—two stations to graph and identify (listed in no particular order)

>> Reno, Nevada: pop. 221,000; elev. 1341 m (4400 ft), 39.5°N 119.5°W

>> Walgett, New South Wales, Australia: pop. 2000; elev. 133 m (436 ft), 30°S 148°E

Section 5—Your climate region (optional station exercise)

Example climograph:

Station #1: Salvador (Bahia), Brazil

Total annual rainfall: *183.6 cm (72.3 in.)*

Average annual temperature: *24.8°C (76.6°F)*

Annual temperature range: *3.4 C° (6.1 F°)*

Distribution of temperature during the year: *Consistently warm; daily range greater than seasonal range*

Distribution of precipitation during the year: *Each month receives 7.4 to 29.7 cm of rain. March through July are wetter than the rest of the year.*

Principal atmospheric lifting mechanism(s): *Convergent lifting during wetter season with convectional lifting the rest of the year*

Distribution of potential evapotranspiration during the year: *Consistent, never below 8.6 cm (3.4 in.)*

▲Figure 17.2

	Jan	Feb	Mar	Apr	May	Jun	Jul	Aug	Sep	Oct	Nov	Dec	Annual
Temperature °C	26.0	26.3	26.3	25.8	24.8	23.8	23.0	22.9	23.6	24.5	25.1	25.6	24.8
(°F)	(78.8)	(79.3)	(79.3)	(78.4)	(76.6)	(74.8)	(73.4)	(73.2)	(74.5)	(76.1)	(77.2)	(78.1)	(76.6)
PRECIP cm	7.4	7.9	16.3	29.0	29.7	19.6	20.6	11.2	8.4	9.4	14.2	9.9	183.6
(in.)	(2.9)	(3.1)	(6.4)	(11.4)	(11.7)	(7.7)	(8.1)	(4.4)	(3.3)	(3.7)	(5.6)	(3.9)	(72.3)
POTET cm	13.7	12.4	13.5	11.9	10.7	9.1	8.6	8.6	9.4	11.2	11.9	13.2	134.4
(in.)	(5.4)	(4.9)	(5.3)	(4.7)	(4.2)	(3.6)	(3.4)	(3.4)	(3.7)	(4.4)	(4.7)	(5.2)	(52.9)

What are the main climatic influences for this station (air pressure, air mass sources, degree of continentality, temperature of ocean currents)? *Station is under the ITCZ year round, with mT air masses, low continentality as shown by the extremely small temperature range, warm currents offshore.*

Köppen climate classification symbol: *Af*; name: *Tropical rain forest;* explanation for this determination: *All months above 18°C (64.4°F) and all months receive PRECIP over 6 cm (2.4 in.).*

Representative biome (terrestrial ecosystem) characteristic of region: *ETR, equatorial and tropical rain forest;* characteristic vegetation: *evergreen rain forest*

Station #2:

Total annual rainfall:

Average annual temperature:

Annual temperature range:

Distribution of temperature during the year:

Distribution of precipitation during the year:

Principal atmospheric lifting mechanism(s):

Distribution of potential evapotranspiration during the year:

▲ Figure 17.2a

	Jan	Feb	Mar	Apr	May	Jun	Jul	Aug	Sep	Oct	Nov	Dec	Annual
Temperature °C	−10.0	−9.4	−3.3	5.6	13.3	18.3	22.3	19.4	15.0	8.3	0.6	−6.7	6.0
(°F)	(14.0)	(15.1)	(26.1)	(42.1)	(55.9)	(64.9)	(72.1)	(66.9)	(59.0)	(46.9)	(33.1)	(19.9)	(42.8)
PRECIP cm	9.6	7.7	8.8	6.6	8.0	8.7	9.5	8.8	9.3	8.7	9.0	9.1	103.8
(in.)	(3.8)	(3.0)	(3.5)	(2.6)	(3.1)	(3.4)	(3.7)	(3.5)	(3.7)	(3.4)	(3.5)	(3.6)	(40.8)
POTET cm	0	0	0	2.7	8.1	11.9	13.9	12.1	7.7	3.7	0.2	0	60.3
(in.)	(0)	(0)	(0)	(1.1)	(3.2)	(4.7)	(5.5)	(4.8)	(3.0)	(1.5)	(0.1)	(0)	(23.7)

What are the main climatic influences for this station (air pressure, air mass sources, degree of continentality, temperature of ocean currents)?

Köppen climate classification symbol: ; name: ;
explanation for this determination:

Representative biome (terrestrial ecosystem) characteristic of region: ; characteristic vegetation:

Station #3:

Total annual rainfall:

Average annual temperature:

Annual temperature range:

Distribution of temperature during the year:

Distribution of precipitation during the year:

Principal atmospheric lifting mechanism(s):

Distribution of potential evapotranspiration during the year:

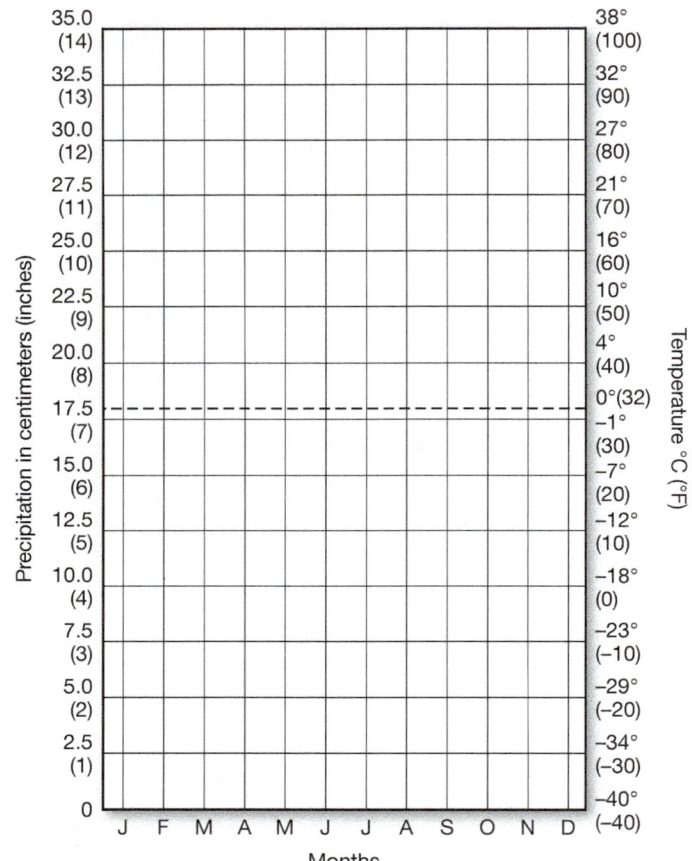

▲ **Figure 17.2b**

	Jan	Feb	Mar	Apr	May	Jun	Jul	Aug	Sep	Oct	Nov	Dec	Annual
Temperature °C	−47.2	−42.9	−34.2	−15.4	1.4	11.6	14.8	10.9	1.6	−16.2	−35.0	−44.0	−16.3
(°F)	(−53.0)	(−45.2)	(−29.6)	(4.3)	(34.5)	(52.9)	(58.6)	(51.6)	(34.9)	(2.8)	(−31.0)	(−47.2)	(2.7)
PRECIP cm	0.8	0.5	0.5	0.3	1.0	3.3	4.1	3.8	2.0	1.3	1.0	0.8	19.3
(in.)	(0.3)	(0.2)	(0.2)	(0.1)	(0.4)	(1.3)	(1.6)	(1.5)	(0.8)	(0.5)	(0.4)	(0.3)	(7.6)
POTET cm	0	0	0	0	3.0	13.5	15.7	10.9	2.3	0	0	0	45.5
(in.)	(0)	(0)	(0)	(0)	(1.2)	(5.3)	(6.2)	(4.3)	(0.9)	(0)	(0)	(0)	(17.9)

What are the main climatic influences for this station (air pressure, air mass sources, degree of continentality, temperature of ocean currents)?

Köppen climate classification symbol: ; name: ; explanation for this determination:

Representative biome (terrestrial ecosystem) characteristic of region: ; characteristic vegetation:

Station #4:

Total annual rainfall:

Average annual temperature:

Annual temperature range:

Distribution of temperature during the year:

Distribution of precipitation during the year:

Principal atmospheric lifting mechanism(s):

Distribution of potential evapotranspiration during the year:

▲ **Figure 17.2c**

	Jan	Feb	Mar	Apr	May	Jun	Jul	Aug	Sep	Oct	Nov	Dec	Annual
Temperature °C	13.3	14.4	17.2	21.1	24.4	27.8	28.3	28.3	26.7	22.8	16.7	13.9	21.1
(°F)	(56.0)	(58.0)	(63.0)	(70.0)	(76.0)	(82.0)	(83.0)	(83.0)	(80.0)	(73.0)	(62.0)	(57.0)	(70.0)
PRECIP cm	12.2	10.7	16.8	13.7	13.7	14.2	18.0	16.3	14.7	9.4	10.2	11.7	161.3
(in.)	(4.8)	(4.2)	(6.6)	(5.4)	(5.4)	(5.6)	(7.1)	(6.4)	(5.8)	(3.7)	(4.0)	(4.6)	(63.5)
POTET cm	2.2	2.6	4.9	8.4	12.7	16.8	18.0	17.1	13.9	8.8	4.0	2.4	111.8
(in.)	(0.9)	(1.0)	(1.9)	(3.3)	(5.0)	(6.6)	(7.1)	(6.7)	(5.5)	(3.5)	(1.6)	(0.9)	(44.0)

What are the main climatic influences for this station (air pressure, air mass sources, degree of continentality, temperature of ocean currents)?

Köppen climate classification symbol: ; name: ;
explanation for this determination:

Representative biome (terrestrial ecosystem) characteristic of region: ; characteristic vegetation:

Station #5:

Total annual rainfall:

Average annual temperature:

Annual temperature range:

Distribution of temperature during the year:

Distribution of precipitation during the year:

Principal atmospheric lifting mechanism(s):

Distribution of potential evapotranspiration during the year:

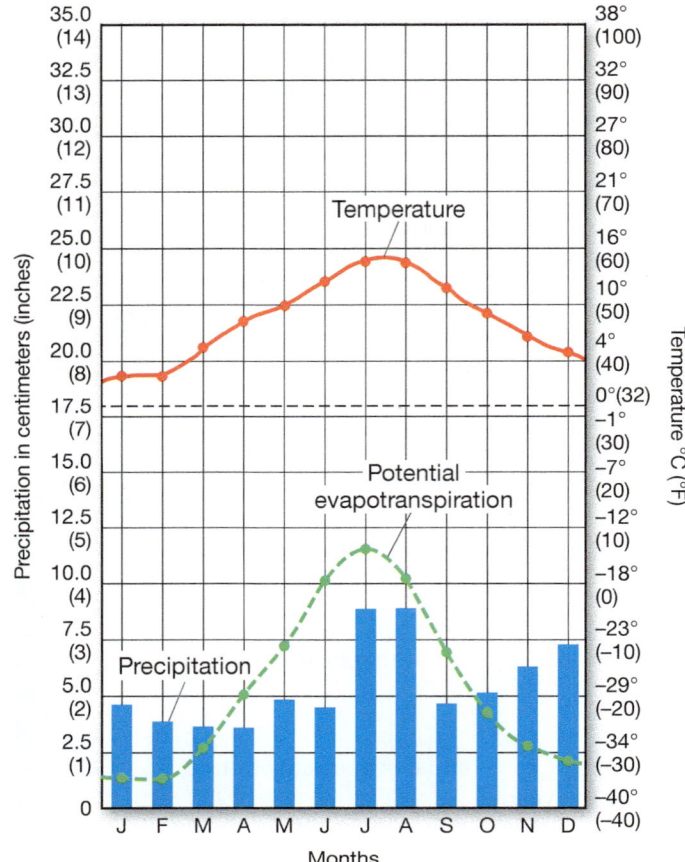

▲ **Figure 17.2d**

	Jan	Feb	Mar	Apr	May	Jun	Jul	Aug	Sep	Oct	Nov	Dec	Annual
Temperature °C	3.0	3.0	5.0	7.6	10.1	12.7	14.7	14.3	12.5	9.7	6.5	4.8	8.7
(°F)	(37.4)	(37.4)	(41.0)	(45.7)	(50.2)	(54.9)	(58.5)	(57.7)	(54.5)	(49.5)	(43.7)	(40.6)	(47.7)
PRECIP cm	4.8	3.6	3.3	3.3	4.8	4.6	8.9	9.1	4.8	5.1	6.1	7.4	65.8
(in.)	(1.9)	(1.4)	(1.3)	(1.3)	(1.9)	(1.8)	(3.5)	(3.6)	(1.9)	(2.0)	(2.4)	(2.9)	(25.9)
POTET cm	1.3	1.3	2.8	4.8	7.6	10.2	11.7	10.2	7.4	4.8	2.8	1.8	64.8
(in.)	(0.5)	(0.5)	(1.1)	(1.9)	(3.0)	(4.0)	(4.6)	(4.0)	(2.9)	(1.9)	(1.1)	(0.7)	(25.5)

What are the main climatic influences for this station (air pressure, air mass sources, degree of continentality, temperature of ocean currents)?

Köppen climate classification symbol: ; name: ;
explanation for this determination:

Representative biome (terrestrial ecosystem) characteristic of region: ; characteristic vegetation:

Station #6:

Total annual rainfall:

Average annual temperature:

Annual temperature range:

Distribution of temperature during the year:

Distribution of precipitation during the year:

Principal atmospheric lifting mechanism(s):

Distribution of potential evapotranspiration during the year:

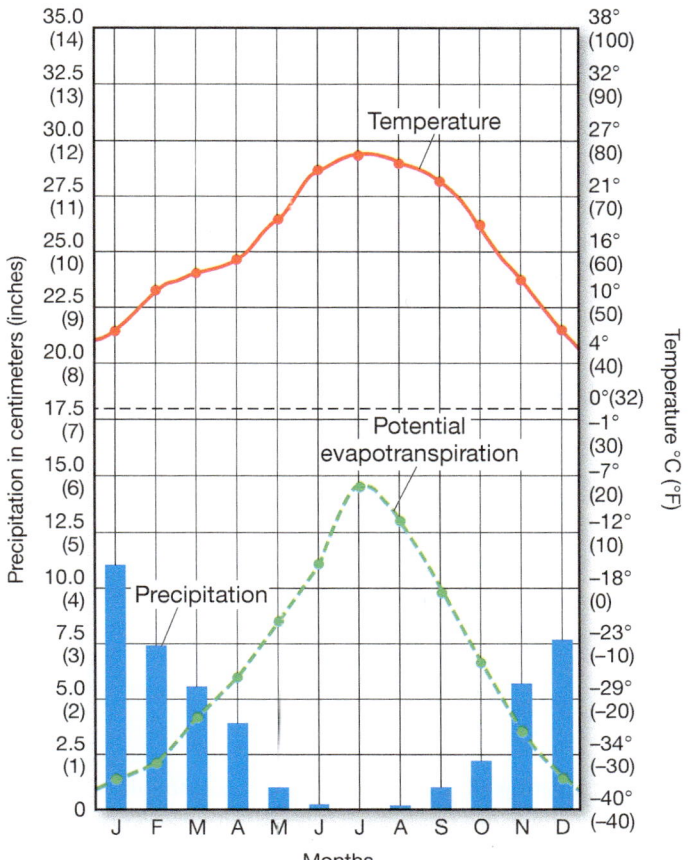

▲ **Figure 17.2e**

	Jan	Feb	Mar	Apr	May	Jun	Jul	Aug	Sep	Oct	Nov	Dec	Annual
Temperature °C	8.4	11.2	12.9	15.6	19.1	22.3	24.8	24.2	22.7	18.5	12.6	8.6	16.8
(°F)	(47.1)	(52.2)	(55.3)	(60.1)	(66.3)	(72.2)	(76.6)	(75.6)	(72.9)	(65.3)	(54.7)	(47.5)	(62.2)
PRECIP cm	10.7	7.4	5.6	3.6	1.0	0.3	0.0	0.3	0.8	2.3	5.8	7.6	45.5
(in.)	(4.2)	(2.9)	(2.2)	(1.4)	(0.4)	(0.1)	(0.0)	(0.1)	(0.3)	(0.9)	(2.3)	(3.0)	(17.9)
POTET cm	1.5	2.3	4.0	6.0	8.5	11.5	14.0	12.7	9.9	6.6	3.0	1.5	81.5
(in.)	(0.6)	(0.9)	(1.6)	(2.4)	(3.3)	(4.5)	(5.5)	(5.0)	(3.9)	(2.6)	(1.2)	(0.6)	(32.1)

What are the main climatic influences for this station (air pressure, air mass sources, degree of continentality, temperature of ocean currents)?

Köppen climate classification symbol: ; name: ;
explanation for this determination:

Representative biome (terrestrial ecosystem) characteristic of region: ; characteristic vegetation:

SECTION 3

Climate Analysis

Temperature, precipitation, and potential evapotranspiration patterns are the elements used in determining climate classification. Factors influencing these patterns help to more fully understand the distribution of climate regions.

Use *Geosystems* to find background information for some of your responses. Answer the following questions about the completed climographs in Figures 17.2a through 17.2e.

New Orleans—Station # _____

1. What effects do latitudinal position, air mass interactions, and ocean currents have in producing the climate of New Orleans? For instance, why is the winter precipitation of New Orleans higher than in more northerly cities?

2. In examining POTET (potential evapotranspiration) values throughout the year for New Orleans and comparing them to PRECIP (precipitation) received, is there an annual moisture surplus or deficit?

Edinburgh—Station # _____

3. Where are the several places in North America that you find this same climatic regime?

4. What controlling factors contribute to the cool summers in Edinburgh?

Sacramento—Station # _____

5. What factors produce the natural summer drought (high POTET and low PRECIP) experienced in Sacramento? (Consider air masses, pressure systems, and ocean currents.)

6. What challenges would natural vegetation face growing in this climate, as opposed to the climate in Montreal or New Orleans?

Montreal—Station # _____

7. Compare the annual range of temperature (between January and July) in Montreal with your analysis of New Orleans. Why the greater range value for Montreal?

8. What contributing factors (mechanisms and conditions) produce the even distribution of precipitation throughout the year in Montreal?

Oymyakon—Station # _____

9. Explain why this station experiences this climate. What are the contributing factors? Why is the summer so hot? Why is the winter so cold? What are two factors that reduce summer and winter PRECIP?

SECTION 4

The Desert Climates

The *B* climates are the only ones that Köppen classified according to the *amount and annual distribution of precipitation*. POTET (potential evapotranspiration) exceeds PRECIP (precipitation) in all parts of *B* climates, creating varying amounts of permanent water deficits that distinguish the different subdivisions within this climatic group. Vegetation is typically xerophytic—that is, drought resistant, waxy, hard leafed, and adapted to aridity and low transpiration loss. In general terms, the major subdivisions are the *BW deserts*, where PRECIP is *less* than one-half of POTET, and the *BS steppes*, where PRECIP is *more* than one-half of POTET.

In an effort to better approximate the dry climates, Köppen developed simple formulas, graphically presented in **Figure 17.3**, to determine the usefulness of rainfall, based on the season in which it falls—whether it falls principally in the summer with a dry winter, or in the winter with a dry summer, or whether the rainfall is evenly distributed throughout the year. Winter rains are the most effective because they fall at a time of lower POTET.

R = mean annual precipitation

PRECIP concentrated in winter
(wettest winter month 3x driest summer month)

R = mean annual precipitation

PRECIP evenly distributed

R = mean annual precipitation

PRECIP concentrated in summer
(wettest summer month 10x driest winter month)

▲ **Figure 17.3** B climate determination

As in Section 2, complete two climographs (**Figures 17.4a** and **17.4b**). Using the climate classification key presented in Table 17.1 and the three graphs in Figure 17.3, determine the climatic classification for each station and the identity of each station—Reno, Nevada, and Walgett, New South Wales. Complete the analysis and information sheet for each of the stations.

Station #7:

Total annual rainfall:

Average annual temperature:

Annual temperature range:

Distribution of temperature during the year:

Which PRECIP regime from Figure 17.3 is this? Which diagram should you use to evaluate this climate?

Principal atmospheric lifting mechanism(s):

Distribution of potential evapotranspiration during the year:

▲ **Figure 17.4a**

	Jan	Feb	Mar	Apr	May	Jun	Jul	Aug	Sep	Oct	Nov	Dec	Annual
Temperature °C	28.1	27.2	24.4	19.7	15.0	11.7	10.8	12.5	19.2	20.6	24.4	27.2	20.0
(°F)	(82.5)	(81.0)	(76.0)	(67.5)	(59.0)	(53.0)	(51.5)	(54.5)	(66.5)	(69.0)	(76.0)	(81.0)	(68.0)
PRECIP cm	5.3	4.8	4.1	3.0	3.8	4.1	3.3	2.8	2.5	3.0	3.8	4.3	45.0
(in.)	(2.1)	(1.9)	(1.6)	(1.2)	(1.5)	(1.6)	(1.3)	(1.1)	(1.0)	(1.2)	(1.5)	(1.7)	(17.7)
POTET cm	17.8	14.5	11.9	6.6	3.6	1.8	1.8	2.5	6.3	8.4	12.7	17.3	104.6
(in.)	(7.0)	(5.7)	(4.7)	(2.6)	(1.4)	(0.7)	(0.7)	(1.0)	(2.5)	(3.3)	(5.0)	(6.8)	(41.2)

What are the main climatic influences for this station (air pressure, air mass sources, degree of continentality, temperature of ocean currents)?

Köppen climate classification symbol: ; name: ; explanation for this determination:

Representative biome (terrestrial ecosystem) characteristic of region: ; characteristic vegetation:

Station #8:

Total annual rainfall:

Average annual temperature:

Annual temperature range:

Distribution of temperature during the year:

Which PRECIP regime from Figure 17.3 is this? Which diagram should you use to evaluate this climate?

Principal atmospheric lifting mechanism(s):

Distribution of potential evapotranspiration during the year:

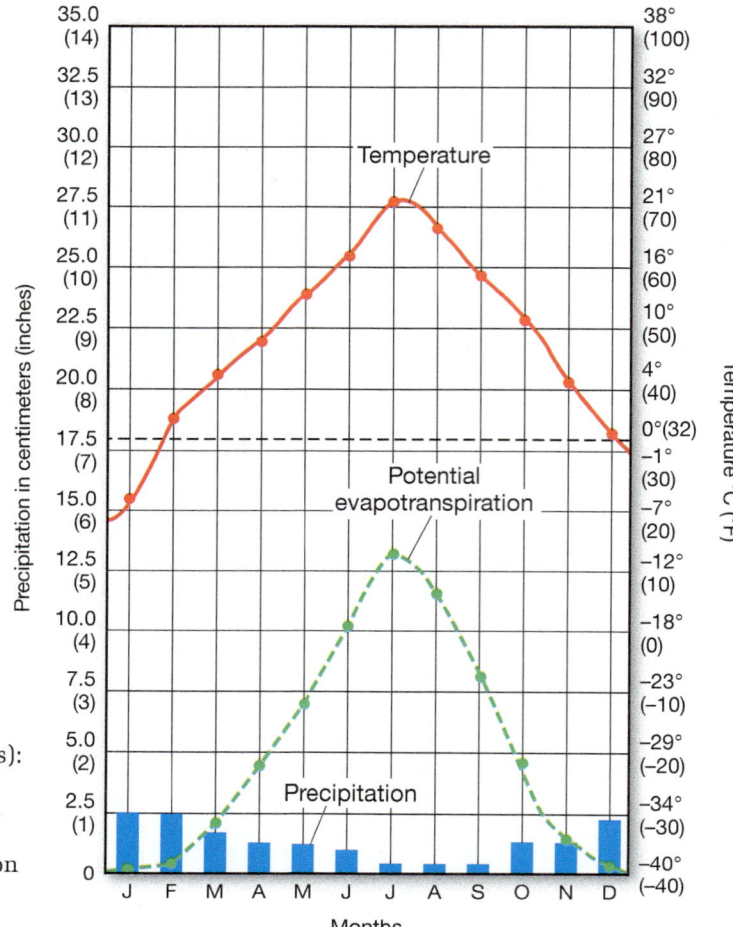

▲ Figure 17.4b

	Jan	Feb	Mar	Apr	May	Jun	Jul	Aug	Sep	Oct	Nov	Dec	Annual
Temperature °C	−0.6	2.2	5.0	8.9	12.8	16.7	21.1	19.4	15.6	10.6	4.4	0.6	10.0
(°F)	(31.0)	(36.0)	(41.0)	(48.0)	(55.0)	(62.0)	(71.7)	(67.0)	(60.0)	(51.0)	(40.0)	(33.0)	(50.0)
PRECIP cm	2.5	2.5	1.8	1.3	1.3	1.0	0.5	0.5	0.5	1.5	1.5	2.3	17.8
(in.)	(1.0)	(1.0)	(0.7)	(0.5)	(0.5)	(0.4)	(0.2)	(0.2)	(0.2)	(0.6)	(0.6)	(0.9)	(7.0)
POTET cm	0	0.7	2.2	4.0	7.1	10.1	13.0	11.7	7.8	4.3	1.7	0.2	62.8
(in.)	(0)	(0.3)	(0.9)	(1.6)	(2.8)	(4.0)	(5.1)	(4.6)	(3.1)	(1.7)	(0.7)	(0.1)	(24.7)

What are the main climatic influences for this station (air pressure, air mass sources, degree of continentality, temperature of ocean currents)?

Köppen climate classification symbol: ; name: ;
explanation for this determination:

Representative biome (terrestrial ecosystem) characteristic of region: characteristic vegetation:

1. Characterize water balance conditions in **Reno, Nevada**, during June, July, and August. (Compare precipitation and potential evapotranspiration for these months.)

2. What air mass lifting mechanism produces most of the precipitation received in Reno?

3. What air masses produce most of the precipitation received in January and February?

4. What is the annual range of temperature (between January and July) for Reno?

   What factors contribute to this value for Reno?

5. **Walgett, N.S.W.**, within a temperate short-grass plain region, is in the Southern Hemisphere (**Station # _____**). What indications of this global location can you determine from its temperature, PRECIP, and POTET data? Explain.

6. What climate is suggested by just evaluating PRECIP and POTET? What climate did you determine using Figure 17.3?

SECTION 5

Your Climate Region (optional station exercise)

Now that you have examined the various climate patterns throughout the world, you can better understand the climate region in which you are located. Your instructor will give you the necessary climate data for your campus or for a nearby climate station. Use the data to complete the climograph, analysis, and information sheet (Figure 17.5). Note the climate station from Sections 2 and 4 that is most like your climate. Review the questions about the climate to better understand your local conditions.

Station:

Latitude:

Longitude:

Elevation:

Population:

Total annual rainfall:

Average annual temperature:

Annual temperature range:

Distribution of temperature during the year:

Distribution of precipitation during the year:

Principal atmospheric lifting mechanism(s):

Distribution of potential evapotranspiration during the year:

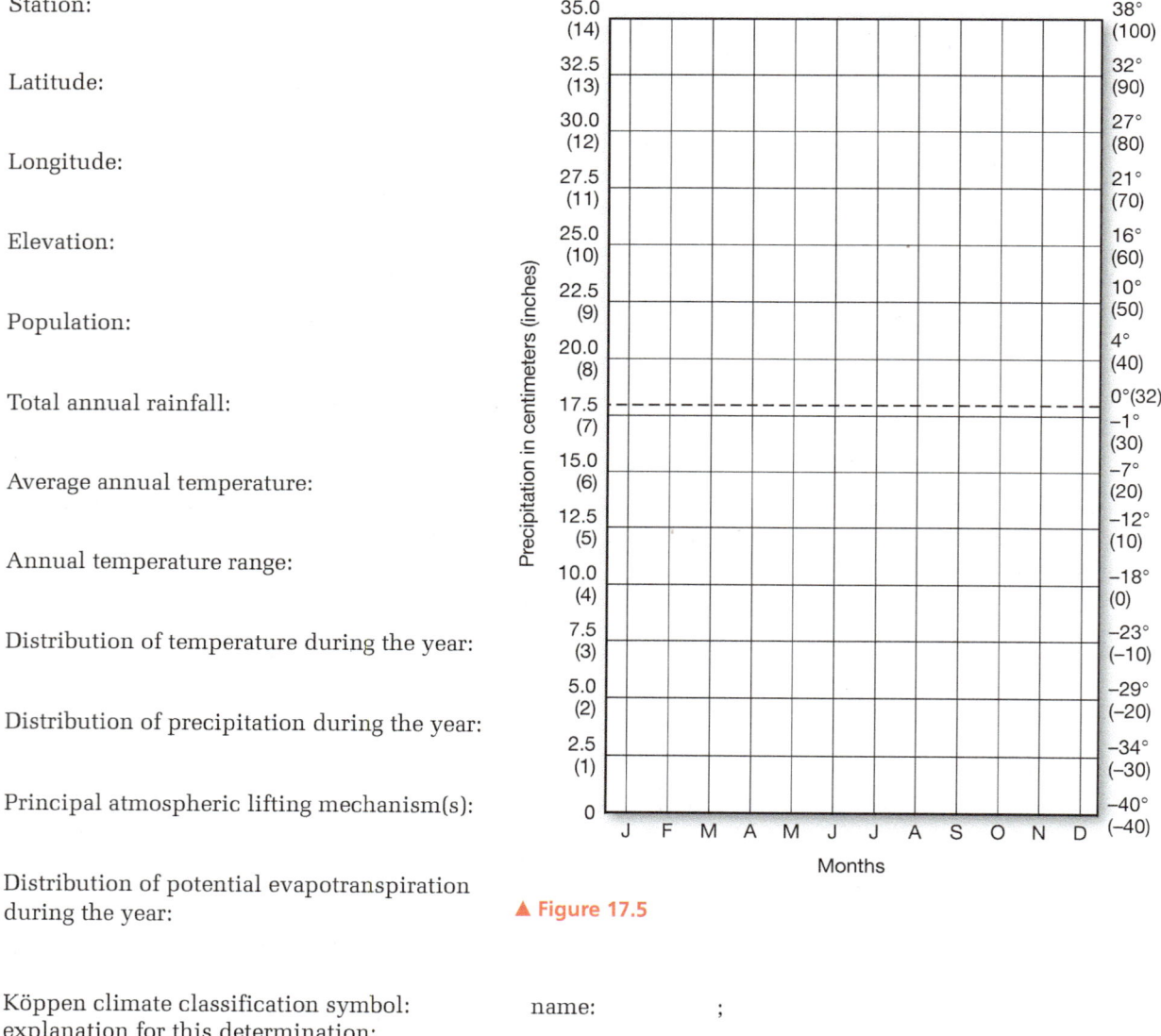

▲ Figure 17.5

Köppen climate classification symbol: name: ;
explanation for this determination:

Mastering Geography

Looking for additional lab prep and review materials? Visit the Study Area in Mastering Geography to enhance your geographic literacy and spatial reasoning skills by accessing MapMaster™ interactive maps, videos, *Mobile Field Trips*, *Project Condor* Quadcopter Videos, *In the News* articles, flashcards, web links, self-study quizzes, and other resources. Contact your professor to find out how you can access Mastering Geography.

Name: _____   Laboratory Section: _____

Date: _____   Score/Grade: _____

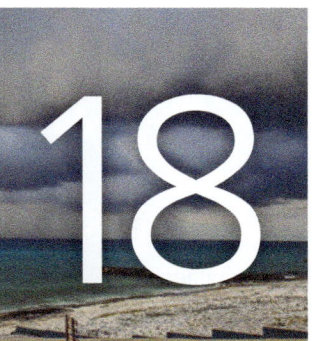

LAB EXERCISE

18

Climate Change

Climate change, a significant and lasting change in weather patterns, ranging from decades to millions of years, is one of the most critical issues facing humankind in the twenty-first century. It is also an integral part of physical geography and Earth systems science. Three key elements of climate change science are the study of past climates, the measurement of current climatic changes, and the forecasting and modeling of future climate scenarios—all of which we will examine in this lab.

The physical evidence for global climate change is extensive and observable—record-breaking global average temperatures for air, land surfaces, lakes, and oceans; ice losses from mountain glaciers and the Greenland and Antarctic ice sheets; declining soil-moisture conditions and resultant effects on crop yields; changing distributions of plants and animals; and the pervasive impact of global sea-level rise, which threatens coastal populations and development worldwide. These are only a few of the many complex and far-reaching issues that climate change science must address to understand and mitigate the environmental changes ahead.

Although climate has fluctuated naturally over Earth's long history, scientists agree that modern climate change is resulting from human activities that produce greenhouse gases. The consensus is overwhelming and includes every professional scientific society, association, and council in the United States and throughout the rest of the world. The observed changes in global climate are happening at a pace much faster than that seen in the historical climate records or in the climate reconstructions that now extend millions of years into the past.

In this lab exercise, we explore the record of past climates, examine the present changes, and look at what changes the future might bring. We will analyze data from **ice cores**, analyze carbon dioxide levels and temperatures, and classify possible future climates. Lab Exercise 18 features three sections.

Key Terms and Concepts

climate change
ice core

isotope analysis
paleoclimatology

proxy method

KEY**LEARNING**concepts

After completion of this lab, you should be able to:

- **Identify** patterns of CO_2 and temperature in ice core data and *evaluate* whether past climate changes have been global or hemispheric.
- **Analyze** the relationship between CO_2 and temperature from the beginning of the Industrial Revolution to the present.
- **Classify** future climate data using the Köppen climate classification system.

Materials/Sources Needed

biomes chart in Exercise 29
color pencils
calculator

spreadsheet program (optional)
Internet access (optional)

Lab Exercises and Activities

SECTION 1

Paleoclimate

To understand present climate changes, we need to understand the climates of the past, especially climates that occurred before record-keeping with standardized instruments began almost 140 years ago. Clues to past climates are stored in a variety of environments on Earth, including gas bubbles in glacial ice, fossil marine life in ocean-bottom sediments, fossil pollen from ancient plants, growth rings in trees, and corals. Scientists extract and analyze cores to establish a chronology of environmental conditions over time periods of thousands or millions of years.

The study of Earth's past climates is the science of **paleoclimatology**, which tells us that Earth's climate has fluctuated over millions of years. To learn about past climates, scientists use **proxy methods** instead of direct measurements. A *climate proxy* is a piece of information from the natural environment that can be used to understand processes that extend back further than our present instrumentation allows. For example, a drier year may reduce the rate of tree growth, resulting in a thin tree ring. The widths of tree rings can tell us about climatic conditions that occurred thousands of years before weather record keeping began. By analyzing evidence from proxy sources, scientists are able to reconstruct climate in ways not possible using the record of firsthand scientific measurements over the past 140 years or so. Some paleoclimatology techniques yield long-term records that span hundreds of thousands to millions of years. Such records come from cores drilled into ocean-bottom sediments or into the thickest ice sheets on Earth. Once cores are extracted, layers containing fossils, air bubbles, particulates, and other materials provide information about past climates.

One basis for long-term climate reconstruction is **isotope analysis**, a technique that uses the atomic structure of chemical elements, specifically the relative amounts of their isotopes, to reveal the chemical composition and temperature of past oceans and ice masses. Oxygen-16, or "lighter" ^{16}O, is the most common isotope found in nature. Oxygen-18, or "heavier" ^{18}O, comprises only about 0.20% of total oxygen atoms. During periods of colder temperatures, more "light" oxygen is locked up in snow

and ice in the polar regions, and "heavy" oxygen concentrations are highest in the oceans.

In the cold regions of the world, snow accumulates seasonally in layers, and in regions where snow is permanent on the landscape, these layers of snow eventually form glacial ice. The world's largest accumulations of glacial ice occur in Greenland and Antarctica, and it is from these ice sheets that scientists have extracted ice cores to reconstruct climate. Such ice cores, drilled thousands of meters deep into the thickest part of an ice sheet, provide a shorter but more detailed record of climate than ocean sediment cores, presently pushing the record back 800,000 years.

Within a core, any given year's accumulation consists of a layer of winter ice and a layer of summer ice, each differing in chemistry and texture. Scientists use oxygen isotope ratios to correlate these layers with environmental temperature conditions. Relatively lower ratios of $^{18}O/^{16}O$ indicate warmer conditions, and higher $^{18}O/^{16}O$ ratios indicate colder temperatures. Therefore, the oxygen isotopes in ice cores are a proxy for air temperature above the snow and ice surfaces. Ice cores also reveal information about past atmospheric composition. Within the ice layers, trapped air bubbles reveal concentrations of gases—mainly carbon dioxide and methane—indicative of past environmental conditions at the time the bubbles were sealed into the ice. By analyzing ice cores, we have reconstructed the temperatures and compositions of gases of the past 800,000 years.

Ice core projects in Greenland have produced data spanning more than 120,000 years. In Antarctica, the Dome C ice core (part of the European Project for Ice Coring in Antarctica [EPICA]), completed in 2004, reached a depth of 3270 m (10,729 ft) and produced the longest ice-core record yet: 800,000 years of Earth's past climate history. This record was correlated with a core record of 400,000 years from the nearby Vostok Station and matched with ocean sediment core records to provide scientists with a sound reconstruction of climate changes throughout this time period. In 2011, American scientists extracted an ice core from the West Antarctic Ice Sheet (WAIS) that when analyzed will reveal 30,000 years of annual climate history and 68,000 years at resolutions from annual to decadal—a higher time resolution than previous coring projects.

Analyses have shown that the greenhouse gas concentrations lag behind the temperature changes, generally by about 1000 years. This interesting relationship suggests the presence and importance of climate feedbacks.

Examine **Figure 18.1**. This EPICA ice core is 3270 m (10,729 ft) long. So far, scientists have analyzed the top 3189.45 m (10,462 ft), revealing the past 801,588 years of climate. The figure shows CO_2 in parts per million (ppm) and temperature anomalies in Celsius degrees relative to the average for the past 1000 years.

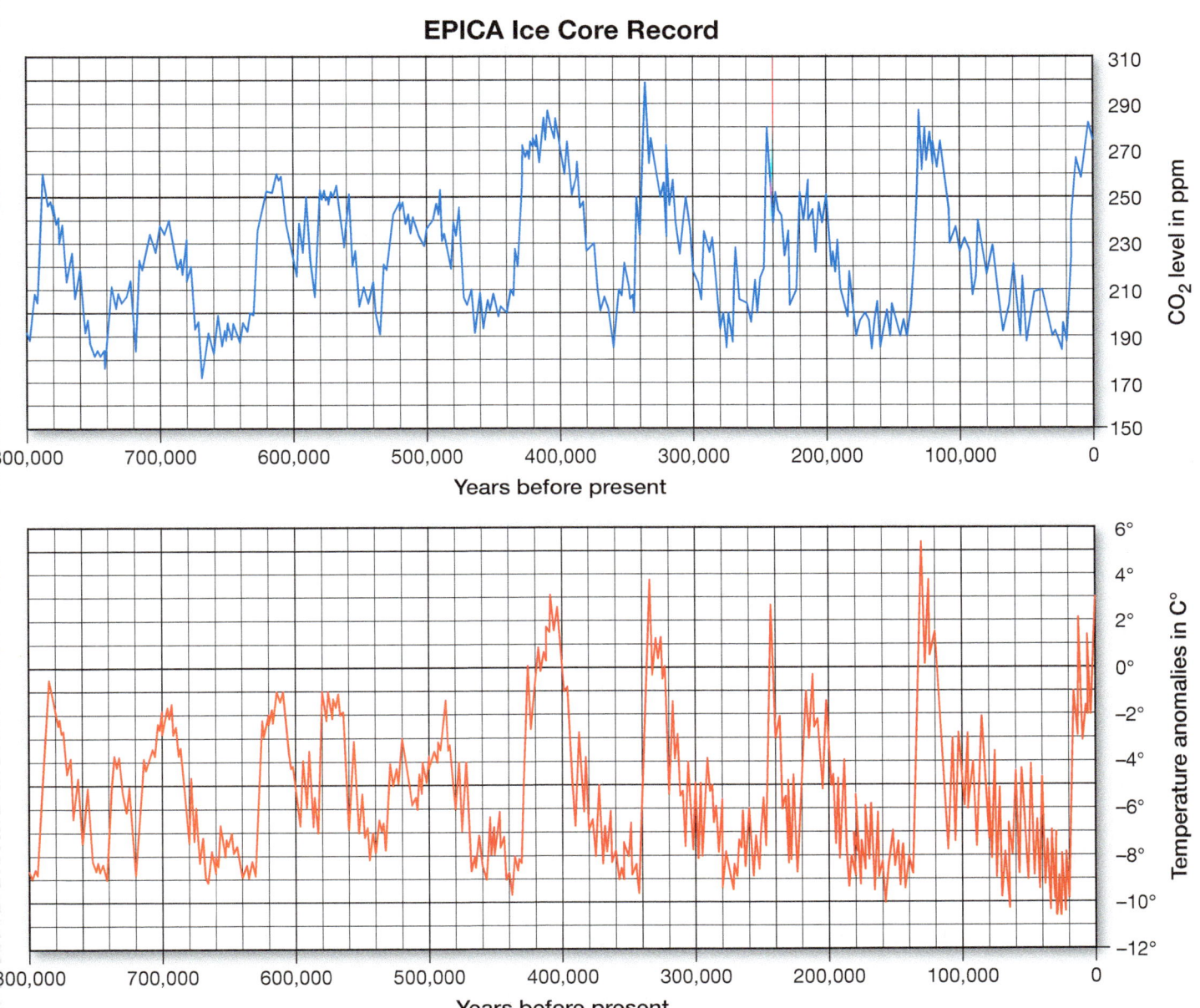

▲ **Figure 18.1** EPICA ice core record

1. How many years of data are displayed from the EPICA ice core?

2. What are the maximum and minimum temperature anomalies recorded? What is the range of temperature anomalies over this record?

3. How many interglacials (times where the temperature deviation reached 0 C° or greater) do you detect over the past 450,000 years? What is the average time between interglacials?

4. How much has CO_2 varied in the EPICA record? What were the highest and lowest values? What was the range of CO_2 values?

5. How many times has CO_2 exceeded 270 ppm in the past 450,000 years?

6. What is the period from peak to peak with regard to CO_2 and for temperature for the past 450,000 years?

7. Interglacial–glacial period climate cycles are often referred to as appearing saw-toothed, with an abrupt change followed by a more gradual change. When do we see an abrupt change: when going from interglacial to glacial times or when going from glacial to interglacial times? When do we see a gradual change?

Examine **Figure 18.2**. The Vostok ice core is 3310 m (10,860 ft) long. The full record extends back to 422,000 years before the present. The figure shows CO_2 in parts per million (ppm) and temperature anomalies in Celsius degrees.

1. How many years of data are displayed from the Vostok ice core?

2. What are the maximum and minimum temperature anomalies recorded? What is the range of temperature anomalies in the Vostok record?

3. How many interglacials (times where the temperature deviation reached 0 C° or greater) do you detect over the past 450,000 years? What is the average time between interglacials?

4. How much has CO_2 varied in the Vostok record? What were the highest and lowest values? What was the range of CO_2 values?

5. What is the period from peak to peak with regard to CO_2 and for temperature for the Vostok record? How does this compare with the period (time from peak to peak) of the EPICA record?

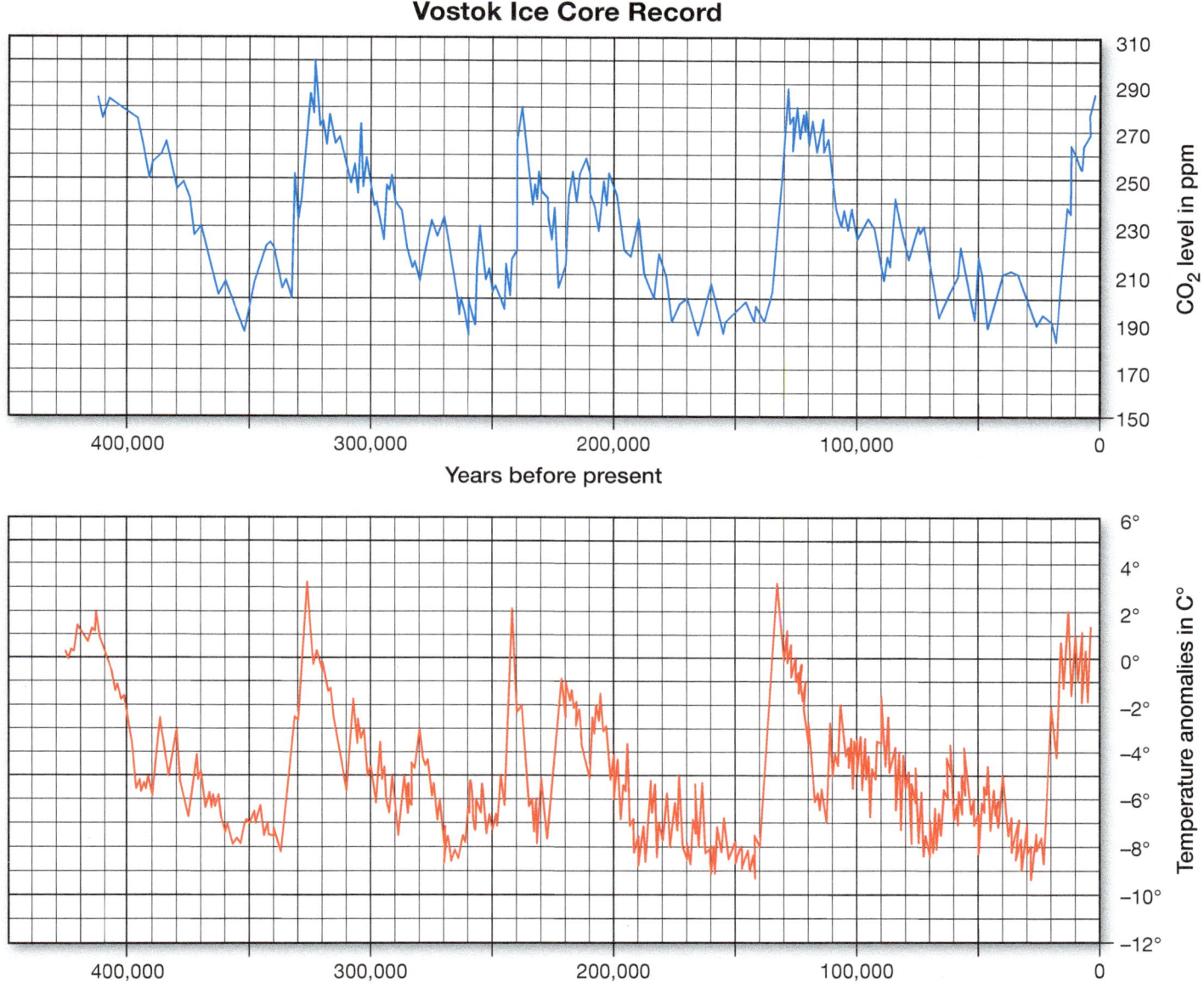

Vostok Ice Core Record

▲ **Figure 18.2** Vostok ice core record

6. Interglacial–glacial period climate cycles are often referred to as appearing saw-toothed, with an abrupt change followed by a more gradual change. When do we see an abrupt change: when going from interglacial to glacial times or when going from glacial to interglacial times? When do we see a gradual change?

7. Why is it important to compare records from different locations? What other locations would you examine to study climate?

SECTION 2

Modern Climate

Carbon dioxide is a natural by-product of life processes, and it is an important component of the atmosphere. Although it accounts for only a small percentage of the atmosphere, about 0.04%, it plays a vital role in maintaining habitable global temperatures. Without CO_2 in the atmosphere, the temperature of Earth would be 15 to 33 C° cooler than it is, with Earth's average temperature around the freezing point of water. As humans have increased carbon dioxide levels in the atmosphere, global temperatures have also increased. The Dome C ice core shows that present levels of carbon dioxide, methane, and nitrous oxides are higher in the present decade than in the rest of the 800,000-year record! These are remarkable times in which we live. The year 2016 was the warmest ever recorded, 0.99°C warmer than the 1951–1980 average. And 2016 was the first time three consecutive warmest years have occurred since the current warming trend began in the 1970s, making it the third record-breaking year in a row. Sixteen of the 17 warmest years ever recorded have occurred since 2000. The last month that Earth saw a cooler than normal month was February 1985. Since 1880, in the Northern Hemisphere the period from 2000 to 2010 was the warmest decade. Data from long-term climate reconstructions of temperature point to the present time as being the warmest in the last 120,000 years. The Intergovernmental Panel on Climate Change (IPCC), through five major reports, including the 2014 Fifth Assessment Report (AR5), confirms that global warming is occurring. The IPCC AR5 reached a consensus, in part stating:

> AR4 concluded that warming of the climate system is unequivocal. New observations, longer data sets, and more paleoclimate information give further support for this conclusion. Confidence is stronger that many changes, that are observed consistently across components of the climate system, are significant, unusual or unprecedented on time scales of decades to many hundreds of thousands of years.*

The CO_2 data for this exercise comes from the Mauna Loa observatory on the Big Island of Hawai'i and is the longest continuous direct measurement of CO_2. Dr. Charles Keeling began the project in 1958. Pay close attention to the scales of the two data series since they cover different time periods. The CO_2 units are parts per million (ppm), and the temperature values are in degrees Celsius.

Figure 18.3 shows the global annual mean temperature in degrees Celsius from 1880 to 2016,

▲ Figure 18.3 Global annual mean temperature

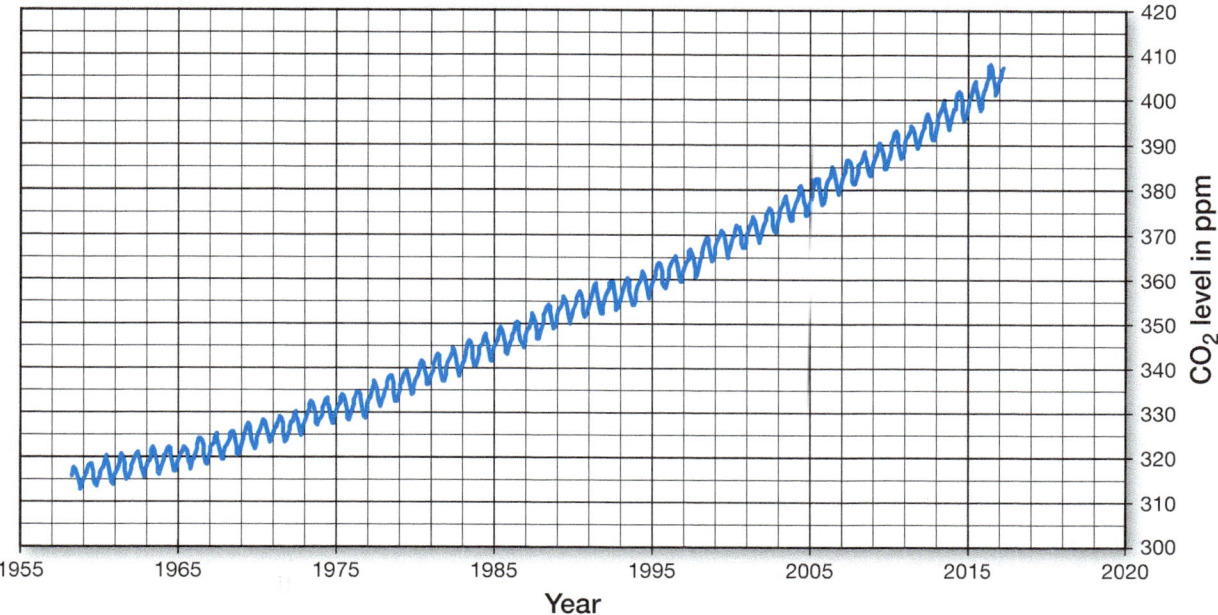

▲ **Figure 18.4** CO_2 1958–2017

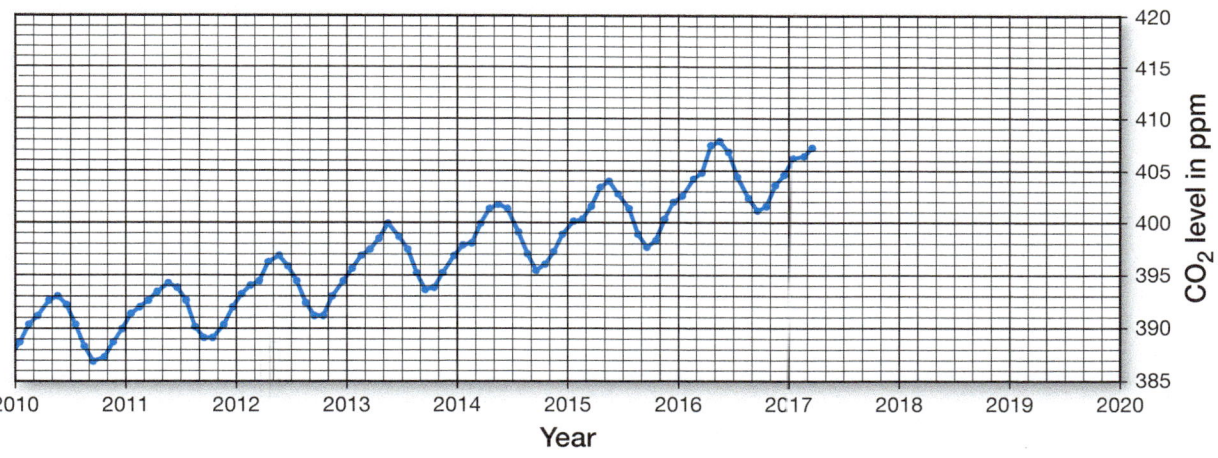

▲ **Figure 18.5** CO_2 2010 – 2017

and **Figure 18.4** shows annual CO_2 levels from 1958 to April 2017, as measured at the Mauna Loa observatory. **Figure 18.5** shows monthly CO_2 values from 2010 to April 2017.

1. What are the general characteristics of the CO_2 data set? What are the maximum and minimum CO_2 concentrations shown in Figure 18.4.? When did the minimum value occur? When did the maximum value occur? What is the general trend in CO_2 values?

2. Compare the CO_2 values from 2006 to 2016 in Figure 18.4 with the CO_2 values on the Vostok plot in Figure 18.2. Are the values for CO_2 levels on the Vostok plot as high as they are now? If so, during which time periods were CO_2 levels as high as they are now? If not, what are the highest values for CO_2 on the Vostok plot?

3. Figure 18.5 shows monthly CO_2 values for 2010–2017. The CO_2 values show a short-term pattern of change. How long is the oscillation?

4. Which months of the year have the highest and lowest values on Figure 18.5?

5. What causes the short-term oscillation of CO_2?

6. What are the annual averages for each of the past 3 years?

7. What is the range from each of the past 3 years to the next year? Compare the range within years to the range between years. For example, compare the changes in CO_2 levels in 2010 to the change from 2010 to 2011.

8. Calculate the average annual increase as a percentage of total CO_2 and in ppm of CO_2 for 1970 to 1980 and for 2006 to 2016.

9. What is the present level of CO_2? What would it be if it were 50% greater?

10. Based on your calculation for the rate of increase from 2006 to 2016, how long do you estimate it will take for the CO_2 level to reach 450 ppm? How long to be 50% higher than the present level?

11. The Kyoto Protocol calls for an overall reduction of greenhouse gases of 5.3% below the 1990 levels. What was the average level of CO_2 in 1990? What would the level of CO_2 be if the Kyoto goal was met? How many ppm below current levels would that be? The "Group of 77" countries favor a 15% reduction below 1990 levels. What would that level be, and how far below present levels is it?

TABLE 18.1 Major volcanic eruptions since 1867					
Date of Eruption	Average Temp Year Before Eruption	Average Temp Year of Eruption	Average Temp 1 Year After Eruption	Average Temp 2 Years After Eruption	Average Temp 3 Years After Eruption
1883 Krakatau					
1902 Soufriere/Pelee/ Santa Maria					
1912 Katmai					
1963 Agung					
1968 Fernandina Island					
1982 El Chichon					
1991 Pinatubo					

12. Dr. James Hansen of NASA has stated that we should set a goal of keeping CO_2 levels below 350 ppm. Which year did we first exceed that level? If we decreased CO_2 levels at the same rate that we have increased them over the past 10 years, how long would it take until CO_2 levels were again below 350 ppm?

In order to examine the effects of volcanic eruptions on climate, we will look at the relationship between temperature and several major eruptions. The dashed line on Figure 18.3 at 1883 represents the eruption of Krakatau. Using colored pencils, draw vertical lines on Figure 18.3 at 1902, 1912, 1963, 1968, 1982, and 1991 to represent each of the above volcanic eruptions.

13. Complete **Table 18.1** for any *three* of the above eruptions. How much of an effect do these eruptions have on global temperatures? How much do they change global temperatures? How many years does it take for the climate to return to a normal temperature after a major eruption?

SECTION 3

Future Climate

Climate regions are not static or fixed but are dynamic and respond to changes in the environment. As our climate changes, crop patterns, as well as natural habitats of plants and animals, shift to maintain preferred temperatures. According to climate models for the midlatitudes, climate regions could shift poleward by 150 to 550 km (90 to 350 mi) during this century. For example, climate conditions in Texas are projected to "migrate" northeast, as shown in **Figure 18.6**, so that the future climate in Illinois will resemble the current climate in east Texas.

Use the climograph (**Figure 18.7**) and the data that follows to plot current mean monthly temperature in blue (*solid-line graph*), projected mean monthly temperature in red (*solid-line graph*), current mean monthly precipitation in blue (*a bar graph*), projected mean monthly precipitation in red (*a bar graph*), current potential evapotranspiration in blue (*a dotted line*), and projected potential evapotranspiration in red (*a dotted line*) for the following station. You may have to color in just the outlines of the precipitation bars so you will be able to see both sets of data on the same climograph. Use Table 17.1 in Lab Exercise 17 to determine the station's current and projected climatic classifications. In addition, use the map of Earth's terrestrial biomes on the outside back cover of this manual to establish the current and projected biomes.

(a) Summer

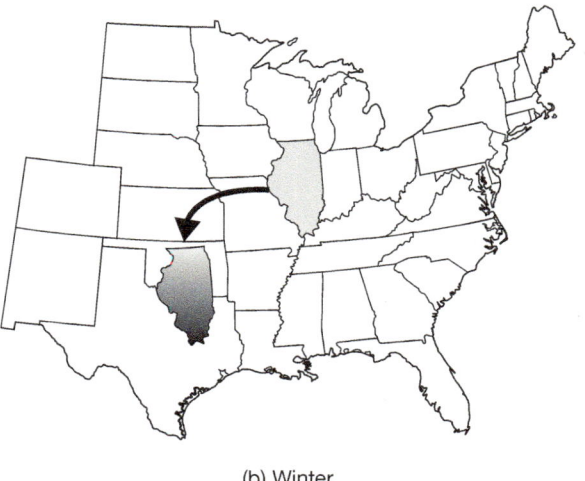
(b) Winter

▲ **Figure 18.6** Migrating summer and winter Illinois climates from 2005 to 2090. These climatic projections are based on moderate CO_2 emission scenarios, which are much lower than current CO_2 emissions. After "Climate Change Projections for the United States Midwest" by D. Wuebbles and K. Hayhoe, *Mitigation Strategies for Global Change 9*: 335–363, 2004. Kluwer Academic Publishers.

1. **Chicago, Illinois**
   Latitude 41.81°N
   Longitude 87.68°W
   Elevation 182 m
   Population 2,714,856
   Current and projected total annual rainfall:

   Current and projected average annual
   temperature:

   Current and projected annual temperature
   range:

   Current and projected distribution of
   temperature during the year:

   Current and projected distribution of precipitation during the year:

   Current and projected distribution of
   potential evapotranspiration during the
   year:

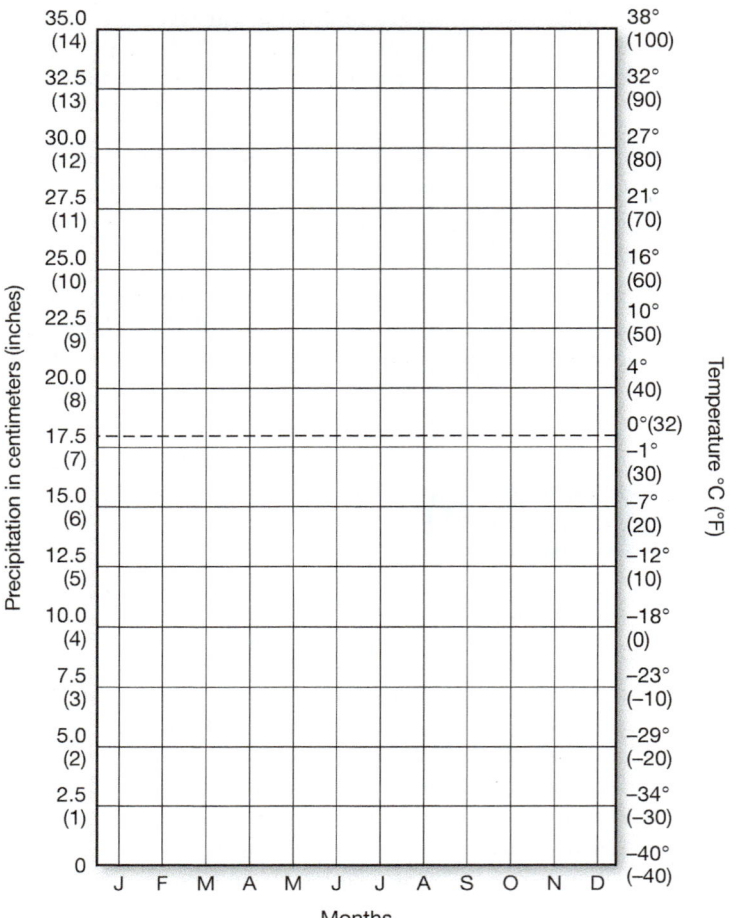

▲ **Figure 18.7**

Current	Jan	Feb	Mar	Apr	May	Jun	Jul	Aug	Sep	Oct	Nov	Dec	Annual
Temperature °C	−6.0	−2.1	2.8	9.4	16.0	21.2	24.1	23.7	19.4	13.1	5.5	−2.3	10.4
(°F)	(21.2)	(28.2)	(37.0)	(48.9)	(60.8)	(70.2)	(75.4)	(74.7)	(66.9)	(55.6)	(41.9)	(27.9)	(50.7)
PRECIP cm	4.7	4.3	5.9	10.1	8.5	10.4	8.5	7.3	9.5	7.6	7.2	6.2	90.2
(in.)	(1.9)	(1.7)	(2.3)	(4.0)	(3.3)	(4.1)	(3.3)	(2.9)	(3.7)	(3.0)	(2.8)	(2.4)	(35.5)
POTET cm	0.0	0.0	0.9	3.9	8.4	12.8	14.7	13.3	9.0	5.0	0.8	0.0	68.8
(in.)	(0.0)	(0.0)	(0.4)	(1.5)	(3.3)	(5.0)	(5.8)	(5.2)	(3.5)	(2.0)	(0.3)	(0.0)	(27.1)

Projected	Jan	Feb	Mar	Apr	May	Jun	Jul	Aug	Sep	Oct	Nov	Dec	Annual
Temperature °C	2.9	6.9	13.6	19.1	22.6	31.0	34.1	33.1	28.2	20.0	11.8	4.6	19.0
(°F)	(37.1)	(44.3)	(56.4)	(66.4)	(72.7)	(87.8)	(93.4)	(91.6)	(82.8)	(68.0)	(53.2)	(40.3)	(66.2)
PRECIP cm	6.6	5.4	9.2	14.8	11.6	6.0	7.4	7.7	9.8	8.8	11.5	7.8	106.6
(in.)	(2.6)	(2.1)	(3.6)	(5.8)	(4.6)	(2.4)	(2.9)	(3.0)	(3.9)	(3.5)	(4.5)	(3.1)	(42.0)
POTET cm	0.0	0.5	2.9	7.1	11.5	24.3	29.5	25.6	15.9	6.5	1.7	0.2	125.7
(in.)	(0.0)	(0.2)	(1.1)	(2.8)	(4.5)	(9.6)	(11.6)	(10.1)	(6.3)	(2.6)	(0.7)	(0.1)	(49.5)

2. Current Köppen climate classification symbol: :

   Name:

   Explanation for this determination:

   Projected Köppen climate classification symbol: :

   Name:

   Explanation for this determination:

3. Compare the position of Chicago, Illinois, in Figure 18.6(a) to the world climate map inside the back cover of this lab manual. What climate classification border is closest to Chicago, Illinois, under current and projected conditions? Describe its future relations to conditions on this map.

   Present location:

   Projected location:

4. Using the world biome map outside the back cover of this lab manual or the one in your textbook, determine:

   Current representative biome (terrestrial ecosystem) characteristic of the region: ;

   Characteristic vegetation:

   Projected representative biome (terrestrial ecosystem) characteristic of the region: ;

   Characteristic vegetation:

5. Evaluate Chicago's current annual March–May, June–August, and September–November water balances. What is the annual surplus or deficit? What are Chicago's projected annual March–May, June–August, and September–November water balances? What is the annual surplus or deficit? Which months have the greatest projected surplus? Which months have the greatest projected deficits?

6. Under moderate climate change scenarios, summer heat index temperatures in the twenty-first century are expected to rise in the American Midwest in a range between 2.7 C° and 13.8 C° (5 F° and 25 F°). A heat wave, similar to the 2003 heat wave that resulted in more than 70,000 deaths across Europe, could result in a tenfold increase in heat-related deaths in Chicago. How do you think this will affect air conditioning demand and related costs? What are your concerns, and what actions would you take if you were a Chicago public health official?

7. Go to www.globalchange.gov and find the National Climatic Assessment report for your region. Briefly summarize the impacts for your region on temperature, precipitation, heat waves, agricultural productivity, and any other significant changes. How are patterns of rainfall projected to shift? For example, will they stay the same, or will there be longer periods of no rain followed by heavy rain and flooding?

Mastering Geography

Looking for additional lab prep and review materials? Visit the Study Area in Mastering Geography to enhance your geographic literacy and spatial reasoning skills by accessing MapMaster™ interactive maps, videos, *Mobile Field Trips*, *Project Condor* Quadcopter Videos, *In the News* articles, flashcards, web links, self-study quizzes, and other resources. Contact your professor to find out how you can access Mastering Geography.

Name: _____   Laboratory Section: _____

Date: _____   Score/Grade: _____

Video
Exercise 19
Pre-Lab Video

http://goo.gl/Xae7TW
Scan to view the
Pre-Lab video

LAB EXERCISE

19 Plate Tectonics: Global Patterns

The **plate tectonics** theory was a revolution in twentieth-century Earth science. The past few decades have seen profound breakthroughs in our understanding of how the continents and oceans evolved, why earthquakes and volcanoes occur where they do, and the reasons for the present arrangement and movement of landmasses. One task of physical geography is to explain the *spatial* implications of all this new knowledge and its effect on Earth's surface and society.

As Earth solidified after formation, heavier elements slowly gravitated toward the center, and lighter elements slowly welled upward to the surface, concentrating in the crust. Earth's interior is highly structured, with uneven heating generated by the radioactive decay of unstable elements. The results of this heating and instability are irregular patterns of moving, warping, and breaking of the crust.

In this lab exercise, we work with plate tectonics and the evolution of the present configuration of continents and ocean basins. A reality of plate tectonics is that pieces of Earth's crust are on the move, actively migrating and interacting. Plate interactions are characterized and portrayed in this lab. The motion of the Pacific plate and the formation of the Hawaiian Islands are analyzed. The ages of oceanic crust and the rates of movement of tectonic plates are calculated. Lab Exercise 19 features mobile-friendly QR links to Google Earth virtual tours that fly you to the topographic map location, as well as optional online Google Earth activities with downloadable media and worksheets at mygeoscienceplace.com and the Mastering Geography Study Area.

Key Terms and Concepts

continental drift
convergent boundary
divergent boundary
hot spots
mid-oceanic ridges

orogenesis
Pangaea
plate tectonics
plumes
sea-floor spreading

subduction zone
transform faults
transform boundary

KEY**LEARNING**concepts

After completion of this lab, you should be able to:

- *Analyze* the arrangement of Earth's crustal plates in their former positions.
- *Contrast* types of plate boundaries and *locate* examples of each type as they are at present.
- *Analyze* the movement of the Pacific plate and the formation of the Hawaiian-Emperor seamount chain.

Materials/Sources Needed

colored pencils
length of string
scissors

transparent tape
Google Earth

Lab Exercise and Activities

SECTION 1

Plate Tectonics

Have you ever looked at a world map and noticed that a few of the continental landmasses appear to match shape, like pieces of a jigsaw puzzle—particularly South America and Africa? The incredible reality is that the continental pieces indeed once were fitted together! Continental landmasses not only migrated to their present locations, but they also continue to move at speeds up to 6 cm (2.4 in.) per year.

In 1912, German geophysicist and meteorologist Alfred Wegener publicly presented in a lecture his idea that Earth's landmasses migrate. His book *Origin of the Continents and Oceans* appeared in 1915. Wegener today is regarded as the father of this concept, called **continental drift**. Wegener postulated that all landmasses were united in one supercontinent approximately 225 million years ago, during the Triassic period. This one landmass he called **Pangaea**, meaning "all Earth," shown in **Figure 19.1**. While his hypothesis would eventually

be proven correct, he was unable to provide a mechanism that would explain how the continents migrated.

Wegener's proposal launched a half-century debate leading up to the adoption of plate tectonics as an all-encompassing theory of continental drift, **sea-floor spreading** along submerged **mid-oceanic ridges**, and **subduction zones**. Beginning in the 1950s and continuing through the 1970s, new evidence resulted in the acceptance of Wegner's hypothesis. The main evidence included finding magnetic reversals along spreading centers in the Atlantic, dating the ages of oceanic crust, and conducting seismic studies revealing the locations of plate boundaries. Aided by an avalanche of discoveries, the theory today is universally accepted as an accurate model of the way Earth's surface evolves.

Animation
Breakup of Pangea

https://goo.gl/tuu6KC

Animation
Plate Motions through Time

https://goo.gl/JtNXOt

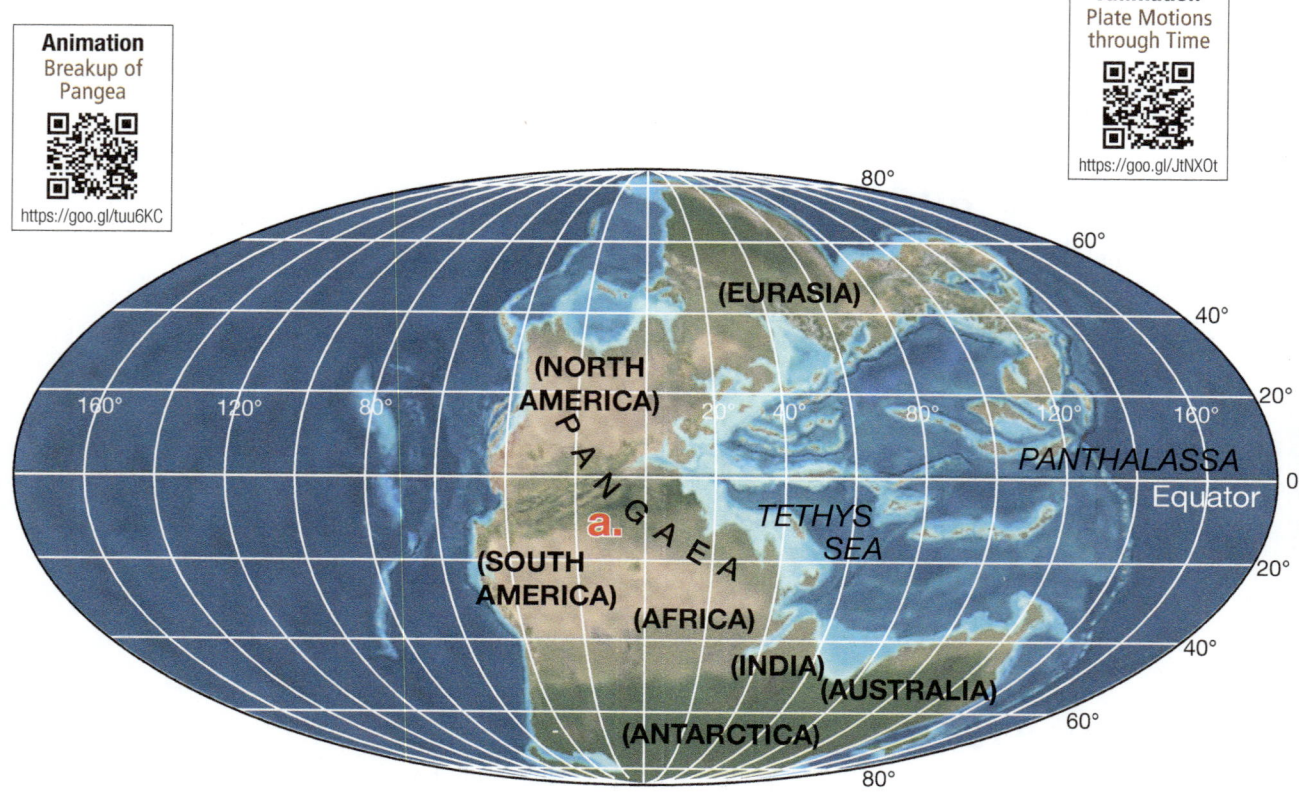

▲ **Figure 19.1** Pangaea 200 to 225 million years ago

The Breakup of Pangaea

Figure 19.1 gives you an opportunity to work with an updated version of Wegener's Pangaea, 225 to 200 million years ago (Triassic–Jurassic periods). Areas of North America, North Africa, the Middle East, and Eurasia were near the equator and therefore were covered by plentiful vegetation. The landmasses focused near the South Pole and attached to the sides of Antarctica were covered by ice. We also see that Africa shared a common connection with both North and South America. The Appalachian Mountains in the eastern United States and the Atlas Mountains of northwestern Africa reflect this common ancestry; they are, in fact, portions of the same mountain range. The Atlantic Ocean did not exist. The only oceans were *Panthalassa* ("all seas"), which would become the Pacific, and the *Tethys Sea* (partly enclosed by the African and the Eurasian plates), which formed the Mediterranean Sea, with trapped portions becoming the present-day Caspian Sea.

Activity and questions about the development of the plate tectonic theory

Figure 19.1 shows Pangaea as it was 200 to 225 million years ago. Alfred Wegener used paleoclimatic evidence to support his idea of plate tectonics. For example, there are tropical regions that have been glaciated, while mid-latitude areas were once tropical swamps. For this section, you will identify locations that had very different climates in the past. For this assignment, we will assume that we are looking at mid-Permian conditions, when global temperatures were similar to current temperatures. You will have to estimate climate and biome locations using latitude and continentality to compare them to present climatic locations. For example, northern India was roughly 35°S and on the east coast of a large landmass. A location with a similar modern climate would be coastal China today, with a humid subtropical climate.

1. For b–d below, write the letter on Figure 19.1 at the appropriate location. The first location (a), has been marked as an example for you. You will need to use Figure 19.1 and the maps of world climates and biomes (found inside the back cover and outside the back cover of this manual) to find and mark:

   a) a location that was a desert (under the subtropical high) but is now a rain forest (tropical). Done for you.

   b) a location that would have been glaciated (Arctic or Antarctic) but is now tropical

   c) a location that was tropical but is now in the midlatitudes

   d) a location that was a rain forest (tropical) but is now a desert (under the subtropical high)

2. What physical clues and evidence prove that these landmasses were together 225 million years ago as Pangaea, and why do scientists think the continents are drifting (continent and sea-floor movements)?

3. If you were a petroleum geologist searching for oil deposits (derived from ancient biomass), why would you want to know the positioning of these landmasses hundreds of millions of years ago? What climate region would you look for?

4. In your opinion, what factors produced the long delay between the time of Wegener's lectures and book and the time of acceptance of his theory?

5. Why did Wegener coin the name "Pangaea"? Explain its meaning.

SECTION 2

Principal Motions of Plates and Plate Boundaries

To visualize the mechanisms that drive Earth's lithospheric plates, refer to **Figure 19.2**. Demonstrated in cross section are sea-floor spreading, upwelling currents, subduction, and plate movements.

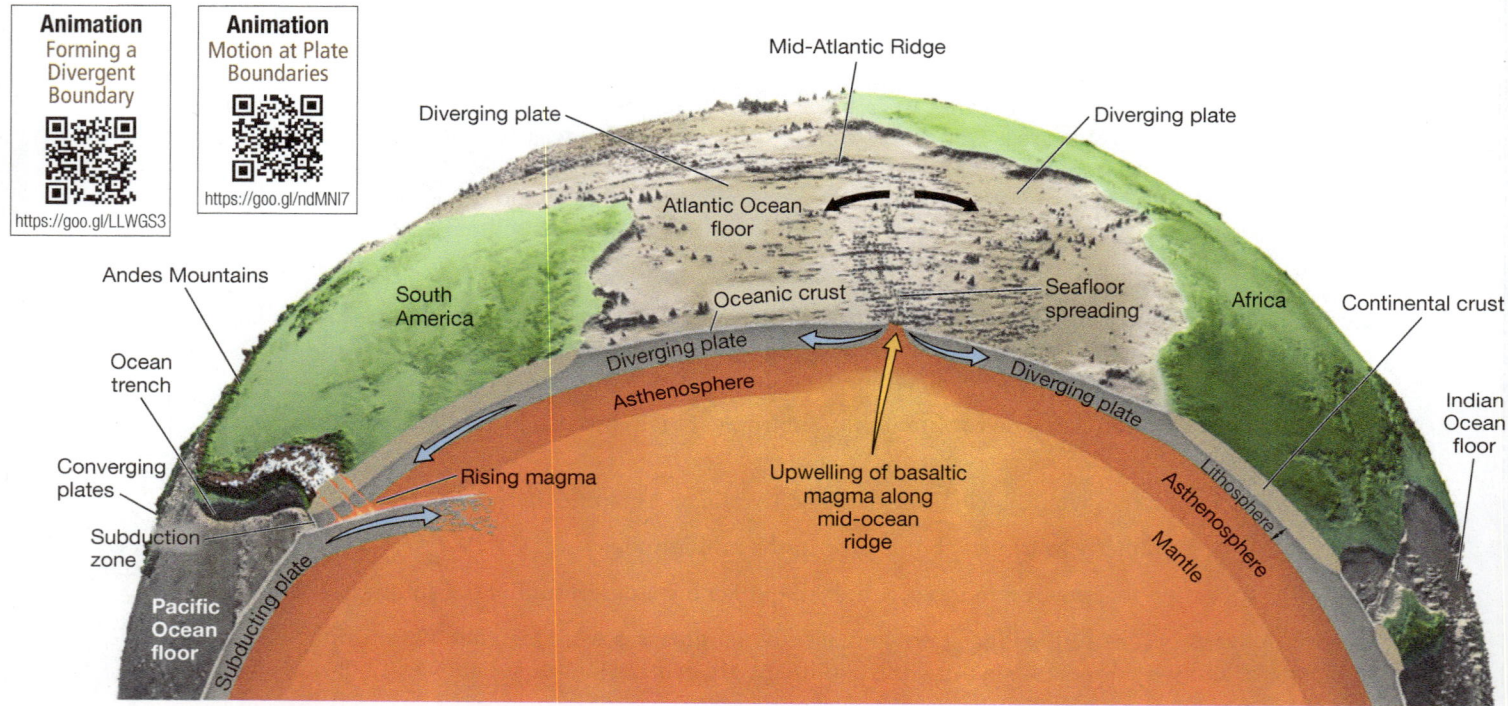

▲ **Figure 19.2** The dynamics of crustal movements.
After Peter J. Wyllie, *The Way the Earth Works*. Copyright © 1976 by John Wiley & Sons, Inc.

1. At the time of Pangaea, South America and Africa were attached, forming a continuous landmass. Using Figure 19.2, describe what transpired in the past 225 million years to position the two continents where they are today.

2. With respect to the movement of crustal plates in the vicinity of the Andes Mountains: Why is the oceanic plate diving beneath South America? How is this related to the **orogenesis** (or formation of mountains) of the Andes? Explain.

Plates and Plate Boundaries

Figure 19.3 shows tectonic features in modern times (the late Cenozoic Era). A spreading center extends down the center of the Atlantic and into the southern Indian Ocean. Subduction trenches are active along the west coasts of Central and South America and throughout the western Pacific Ocean basin. The northern reaches of the Indian plate have underthrust the southern mass of Asia through subduction, forming the Himalayas in the upheaval created by the collision.

Of all the major plates, the Indian traveled the farthest. From approximately 150 million years ago until the present, according to Robert Dietz and John Holden, more than half of the ocean floor was renewed. The three inset blocks demonstrate specific types of plate boundaries: **convergent boundaries**, where two plates are coming together, often leading to the subduction of one plate; **divergent boundaries**, where a plate is spreading apart, such as at sea-floor spreading, mid-ocean ridges); and

Animation
Transform Faults

https://goo.gl/wjND30

Convergent boundary—plates are moving together, producing a subduction zone. Coastal area features mountains, volcanoes, and earthquakes.

Subduction zone

Motion of plate

Spreading ridge offset by transform faults

Transform boundary—plates are sliding past each other along a transform fault.

Divergent boundary—plates are moving apart in areas of seafloor spreading at mid-ocean ridges.

▲ **Figure 19.3** The continents today and types of plate boundary interactions.
Each arrow represents 20 million years of plate movement. The longer arrows indicate that the Pacific and Nazca plates are moving more rapidly than the Atlantic plates.

transform boundaries, where two plates are moving past each other horizontally, for example, the San Andreas Fault. Transform boundaries occur between two plates, while transform faults occur within a plate, for example, the lateral motion between crests of spreading ridges produces transform faults.

Using a physical geography textbook, such as *Geosystems*, locate the plate names, examine the different rates of plate movement, and identify types of plate boundary interactions. Using this same source, complete the following questions.

1. Using colored pencils or colored pens, trace the plate boundaries on Figure 19.3 and color them as follows: convergent boundary–yellow; divergent–red; transform–blue.

2. Of the three types of plate boundary interactions shown in Figure 19.3, characterize the following plate boundaries:

    a) On the ocean floor south of Alaska: _____

    b) Indian Ocean between the Antarctic and Indo-Australian plates: _____

    c) Beneath the Red Sea: _____

    d) Along the west coast of Central America: _____

    e) Through Iceland: _____

    f) Along the northeast coast of the Persian Gulf: _____

    g) In California between the Juan de Fuca plate and the North American plate: _____

3. Which four plates converge on Japan?

4. Mark your location on the map in Figure 19.3. On which plate do you live? (If you live west of the San Andreas fault system in California, take care with your answer.)

5. Google Earth™ activity, San Andreas Fault, CA. For the kmz file and questions, go to mygeoscience-place.com. Then click on the cover of *Applied Physical Geography: Geosystems in the Laboratory*. Or visit the Mastering Geography Study Area.

SECTION 3

Hot Spots and the Hawaiian-Emperor Island Chain

Animation
Hot Spot
Volcano Tracks

https://goo.gl/sOHRDv

A dramatic aspect of plate tectonics is the estimated 50 to 100 hot spots across Earth's surface. These are individual sites of plumes of upwelling material from the mantle. Hot spots occur beneath both oceanic and continental crust and appear to be deeply anchored in the mantle, tending to remain fixed relative to migrating plates.

An example of an isolated hot spot is the one that has formed the Hawaiian-Emperor island chain. The Pacific plate has moved across this hot, upward-erupting plume for over 80 million years, with the resulting string of volcanic islands moving northwestward away from the hot spot. Thus, the age of each island or seamount in the chain increases northwestward from the island of Hawai'i.

The Big Island of Hawai'i actually took less than 1 million years to build to its present stature. Hawai'i Volcanoes National Park is on the island of Hawai'i, where volcanic activity is presently nearly continuous. The youngest island-to-be in the chain is still a *seamount*, a submarine mountain that does not reach the surface. It rises 3350 m (11,000 ft) from the ocean floor but is still 975 m (3200 ft) beneath the ocean surface. Even though this new island will not see the tropical Sun for many years, it is already named Lō'ihi.

The map in **Figure 19.4** shows the ages of each portion or island in the chain called the Hawaiian-Emperor seamount chain.

(b)

(a)

▲ **Figure 19.4** Hawai'i and the Emperor Seamounts.
Ages are shown in millions of years. After D. A. Clague, "Petrology and K-Ar Ages of Dredged Volcanic Rocks from the Western Hawaiian Ridge and the Southern Emperor Seamount Chain," *Geological Society of America Bulletin* 86 (1975): 991.

Using the map in **Figure 19.4**, complete the following questions about the Hawaiian-Emperor island chain. Questions 3 and 4 can be answered using the Google Earth™ activity Hawaii and the Emperor Seamounts. For the kmz file and questions, go to mygeoscience-place.com. Then click on the cover of *Applied Physical Geography: Geosystems in the Laboratory*. Use either a piece of string and the map scale or Google Earth™ to answer the following questions.

1. What is the distance from Midway Island to Lō'ihi, in kilometers? What is the difference between their ages? What has been the rate of movement of the Pacific plate, in centimeters per year, over this time frame?

   [The distance is approximately 2400 km. Midway is 27.7 million years old. Since 1 km equals 100,000 cm, 2400 km is equal to 240,000,000 cm. Since both the distance and age are in the millions, the math to find the rate of movement is 240 (distance in millions of centimeters)/27.7 (age in millions of years) = 8.6 cm/ yr. The conversion from km to centimeters will come in handy in the next section as well.]

2. What is the distance from the Detroit Seamount to the Koko Seamount, in kilometers? What is the difference between their ages? What has been the rate of movement of this region of the Pacific plate, in centimeters per year, over this time frame?

3. What is the distance from the Abbott Seamount to Lō'ihi, in kilometers? What is the difference between their ages? What has been the rate of movement of this region of the Pacific plate, in centimeters per year, over this time frame?

4. What is the distance from the Detroit Seamount through the Daikakuji Seamount to Lō'ihi, in kilometers? What is the difference between their ages? What has been the rate of movement of this region of the Pacific plate, in centimeters per year, over this time frame?

5. Has the rate of movement of the Pacific plate been constant over the lifetime of the Hawaiian-Emperor island chain? Which section has been moving faster?

6. The oldest rock that makes up Lō'ihi is an estimated 400,000 years old. Using the rate of plate movement that you calculated in question 3, how far has it traveled since its first lava cooled?

7. What is the nature of the relative movement of plates and hot spots? Briefly explain which is fixed and which is moving. Also explain how the Hawaiian-Emperor island chain was formed. Part of being a scientist is questioning everything and testing everything. Thinking creatively, what are two possible reasons for the bend in the island chain? Briefly explain how you could test these two ideas.

SECTION 4

Speeds of Plate Movement

Animation
Convection within the Mantle

https://goo.gl/5DqKqc

Condor
Continental Rifting

https://goo.gl/XV3Cho

The key to establishing the theory of continental drift was a better understanding of the sea floor. The sea floor has a remarkable feature: an interconnected worldwide mountain chain, forming a ridge some 64,000 km (40,000 mi) in extent and averaging more than 1000 km (600 mi) in width. In the early 1960s, geophysicist Harry H. Hess proposed sea-floor spreading as the mechanism that builds this mountain chain and drives continental movement. Hess said that these submarine mountain ranges were mid-ocean ridges and the direct result of upwelling flows of magma from hot areas in the upper mantle and asthenosphere and perhaps from the deeper lower mantle. Hess and Robert S. Dietz proposed that old sea floor sinks back into the Earth's mantle at deep ocean trenches and in subduction zones where plates collide.

Today, scientists know that mid-ocean ridges occur at divergent plate boundaries, where plates are moving apart. As the sea floor spreads, magma rises and accumulates in magma chambers beneath the centerline of the ridge. Some of the magma rises and erupts through fractures and small volcanoes along the ridge, forming new sea floor. As the plates continue to move apart, more magma rises from below to fill the gaps. Scientists now think that this upward movement of material beneath an ocean ridge is a result of sea-floor spreading rather than the cause.

Figure 19.5 shows the ages of oceanic crust. The oldest oceanic crust shown on this map is 160–200 million years old. Not shown are parts of the eastern Mediterranean Sea, which has some of the oldest oceanic crust—about 270 million years old.

▲ **Figure 19.5**

In this section you will analyze the ages of tectonic plates and the rates of plate movement. Refer to the legend to find the ages of crust and use the scale bar to find distances. The first transect has been measured, and the age and speed have been calculated for you, but here are the steps. Measure from the middle of the ridge toward Point A. From the middle of the ridge to the far edge of the darkest red band represents 20 million years (my), to the far edge of the next red band represents 40 my, to the far edge of first orange band represents 60 my, to the far edge of the second orange band represents

80 my, and to the far edge of the yellow band represents 100 my, and to the green band represents 120 my. On this transect, the far edge of the 120 my band ends at the beginning of the no data section (represented by the pink areas), so the full age of that band cannot be determined. To be certain of the age we are measuring, we will just measure to the far edge of the 80-100 MY section. Note that the far edges of the oldest bands are uncertain. To obtain an age that we are certain of, measure to the far edge of the oldest full band on that transect. From the ridge to Point C, the transect line crosses

the 0-20 my, 20-40 my, 40-60 my, 60-80 my, 80-100 my, 100-120 my, 120-140 my, 140-160 my, and 160-180 my bands. The oldest edge of the 160-180 my band ends in the no data section, so we are unable to accurately measure the length to the end of the 200 my band and instead we will measure to the edge of the 140-160 my band.

1. Measure along the transect lines from the center of the ridge to the edge of the oldest full band and enter that data into **Table 19.1**.

TABLE 19.1 Plate movement			
Leg	Dist. km	Age my	Speed cm/yr
A to ridge	1500	60	2.5 cm/yr
B to ridge			
C to ridge			
D to ridge			
E to ridge			
F to ridge			
H to edge			

2. Enter the ages of the crust along each transect and calculate the speed of movement for each section in Table 19.1.

3. Are the three sections in the Atlantic that you measured moving at the same speed? Are both sides of the ridge moving at the same speed? Which section is moving the fastest?

4. Compare the speed of the Atlantic sections with the Pacific transect. Which plate is moving more quickly?

5. What types of plate boundaries are on the Pacific and Atlantic shores of South America?

Mastering Geography

Looking for additional lab prep and review materials? Visit the Study Area in Mastering Geography to enhance your geographic literacy and spatial reasoning skills by accessing MapMaster™ interactive maps, videos, *Mobile Field Trips, Project Condor* Quadcopter Videos, *In the News* articles, flashcards, web links, self-study quizzes, and other resources. Contact your professor to find out how you can access Mastering Geography.

Video
Exercise 20
Pre-Lab Video

http://goo.gl/ojS0SH

Scan to view the
Pre-Lab video

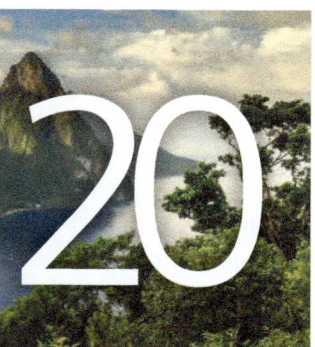

LAB EXERCISE

20 Plate Tectonics: Faulting and Volcanism

Earth's endogenic systems produce flows of heat and material toward the surface to form crust. Ongoing processes produce continental landscapes and oceanic sea-floor crust, sometimes in dramatic episodes. Earth's physical systems move to the front pages whenever an earthquake strikes or after a few of the 50 yearly volcanic eruptions threaten a city, although most of these occurrences are in remote locations or beneath the ocean surface.

We examine processes that construct Earth's surface and create world structural regions. Tectonic processes deform, recycle, and reshape Earth's crust. These processes occur sometimes in dramatic episodes but most often in slow, deliberate motions that build the landscape. Continental crust has been forming throughout most of Earth's 4.6-billion-year existence.

The arrangement of continents and oceans, the origin of mountain ranges, the topography of the land and the seafloor, and the locations of earthquake and volcanic activity are all evidence of our dynamic Earth. Principal seismic and volcanic zones occur along plate boundaries and hot-spot locations, thus linking plate tectonics to the devastation of major earthquakes and local threats of major volcanic eruptions and their attendant potential global climatic impact. We examine the patterns of movement of the San Andreas fault and the morphology of current and ancient volcanic features. We conclude by studying the relationship between volcanic location, magma composition, and volcanic shape.

Lab Exercise 20 features mobile-friendly QR links to Google Earth virtual tours that fly you to the topographic map location, as well as optional online Google Earth activities with downloadable media and worksheets at mygeoscienceplace.com and the Mastering Geography Study Area.

Key Terms and Concepts

asthenosphere	effusive eruption	pyroclastics
cinder cone	elastic-rebound theory	plug dome
crater	explosive eruption	shield volcano
composite volcano	flood basalts	volcano
dikes	lava	

KEY**LEARNING**concepts

After completion of this lab, you should be able to:

- *Identify* typical landscape features associated with faulting.
- *Calculate* rates of movement of the San Andreas fault.
- *Relate* volcanic profiles to volcanic location and chemical composition of magma.
- *Analyze* the Kīlauea Caldera, Hawai'i, and Shiprock, New Mexico, volcanic landscapes.

Materials/Sources Needed

colored pencils	transparent tape
length of string	Google Earth
scissors	

Lab Exercise and Activities

SECTION 1

The San Andreas Fault

Crustal plates do not glide smoothly past one another. Instead, tremendous friction exists along plate boundaries. The specific mechanics of how a fault breaks remain under study, but **elastic-rebound theory** describes the basic process. Generally, two sides along a fault appear to be locked by friction, resisting any movement despite the powerful forces acting on the adjoining pieces of crust. Stress continues to build strain along the *fault plane* surfaces, storing elastic energy like a wound-up spring. When the strain buildup finally exceeds the frictional lock, the two sides of the fault plane lurch into new positions, moving from centimeters to several meters, and release enormous amounts of seismic energy into the surrounding crust. This energy radiates throughout the planet, diminishing with distance but sufficient to register on instruments worldwide.

Pinnacles National Monument and the Neenach formation formed 23 million years ago as a single volcano on the San Andreas fault (**Figure 20.1**). They were together when they formed but have been split apart by the movement of the San Andreas fault.

1. How many kilometers separate Pinnacles and southern portion of the Neenach formation? What is the average speed of movement of this section of the San Andreas fault over the past 23 million years, in centimeters per year?

2. The average rate of motion across the San Andreas fault zone during the past 3 million years is 5.6 cm/yr. How far apart are San Francisco and Palm Springs, in kilometers? How many million years will pass until they are sister cities across a new bay?

The 1906 San Francisco earthquake resulted in a maximum 8.5 m (28 ft) of movement along the San Andreas Fault around Point Reyes.

3. How many years of stress were relieved by the 1906 earthquake, if the speed of movement has been 5.6 cm/yr?

4. If that section of the fault has not moved since 1906, how many centimeters of stress have accumulated, assuming a speed of movement of 5.6 cm/yr?

▲ **Figure 20.1** San Andreas Fault, Pinnacles National Monument, and the Neenach Volcano

SECTION 2

Volcanoes

Volcanic eruptions across the globe remind us of Earth's internal energy. Ongoing eruption activity matches patterns of plate tectonic activity. More than 1300 identifiable land volcanic cones and mountains exist on Earth, although fewer than 600 are active. In an average year, about 50 volcanoes erupt worldwide, varying from modest activity to major explosions.

A **volcano** forms at the end of a central vent or pipe that rises from the asthenosphere and upper mantle through the crust into a volcanic mountain. A **crater**, or circular surface depression, usually forms

at or near the summit. Magma rises and collects in a magma chamber deep below the volcano until conditions are right for an eruption. **Lava** (molten rock), gases, and **pyroclastics**, or *tephra* (pulverized rock and clastic materials of various sizes ejected violently during an eruption), pass through the vent to the surface and build the volcanic landform.

The fact that lava can occur in many different textures and forms accounts for the varied behavior of volcanoes and the different landforms they build. In this section, we look at four volcanic landforms and their origins: shield volcanoes, cinder cones, composite volcanoes, and plug domes.

Volcanoes are of two general types, based on the chemistry and the viscosity of the magma involved. An **effusive eruption** produces a **shield volcano** (such as Kīlauea in Hawai'i) and extensive deposits of **flood basalts**, or plateau basalts (**Figure 20.2c**). Magma of higher viscosity leads to an **explosive eruption** (such as Mt. St. Helens), producing a **composite volcano,** made of multiple flows of lava and ash (**Figure 20.2a**). The chemical content of the magma is related to the location of the volcano and determines its eruptive behavior and overall shape.

Effusive eruptions of low-viscosity fluid basalt flows create shield volcanoes. These volcanoes are gently sloped, gradually rising from the surrounding landscape to a summit crater. The shape is similar in outline to a shield of armor lying face up on the ground, hence the name *shield volcano*. The upper slopes are gently sloped around 5° because the magma is quite hot. However, it cools as it travels, so the lower slopes are a steeper 10°. A related effusive feature is a **cinder cone**, a small, cone-shaped hill usually less than 450 m (1500 ft) high, with a truncated top formed from cinders that accumulate during moderately explosive eruptions (**Figure 20.2b**). Cinder cones are made of pyroclastic material and *scoria* (cindery rock, full of air bubbles) during an initial phase of an effusive eruption. Because they are made of loose material, their slope is quite steep, often with a 34° *angle of repose.* The angle of repose is the slope of a body of loose material.

Volcanic activity near subduction zones produces the infamous explosive volcanoes. Magma produced by the melting of subducted oceanic plate and other materials is thicker (more viscous) than magma that forms effusive volcanoes. It is 50%–75% silica and high in aluminum; consequently, it tends to block the magma conduit. The blockage traps and compresses gases, causing pressure to build for a possible *explosive eruption*, such as the 1980 eruption of Mt. St. Helens.

(a) Cross-section of a composite volcano

(b) Cross-section of a shield volcano

(c) Cross-section of a cinder cone volcano

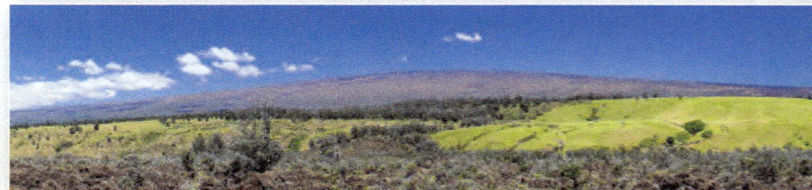

(d) Mauna Loa is a gently sloping shield volcano on the Big Island of Hawai'i

(e) Comparison of the sizes and shapes of Mauna Loa, a shield volcano, and Mount Rainier, a composite volcano. Their different shapes and sizes reflect the different origins and chemical compositions of their magma.

▲ **Figure 20.2** [(d) Alvis Upitis/Photographer's Choice/Getty Images.]

Unlike the volcanoes in Hawai'i Volcanoes National Park, where tourists gather at observation platforms or hike out near the cooling lava flows to watch the relatively calm effusive eruptions, these explosive volcanoes do not invite close inspection and can erupt dangerously with little warning.

Plug dome volcanoes are formed from magma that is high in silica, and very thick. The lava from these volcanoes is too stiff to flow; instead it extrudes from the ground, like toothpaste. A plug dome volcano often has a solid core of cooled lava with flanks made of loose pyroclastics.

Some volcanoes are formed from multiple eruptions—some of ash and some of lava. The term composite volcano describes these explosively formed mountains. They are sometimes called *stratovolcanoes* because they are built up in alternating layers of ash, rock, and lava, but shield volcanoes also can exhibit a stratified structure, so *composite* is the preferred term. Composite volcanoes tend to have steep sides, are more conical in shape than shield volcanoes, and therefore are also known as *composite cones*. If a single summit vent erupts repeatedly, a remarkable symmetry may develop as the mountain grows in size, as demonstrated by Mt. Rainier, Mt. Fuji, and Mt. St. Helens before it erupted in 1980. These volcanoes have a steeper upper section with slopes of up to 30° and lower flanks made of eroded materials with slopes of 5° to 10°.

In this section, you will be calculating the slope and relief of four volcanoes in order to better understand the relationship between their shape and the chemical content of their magma. You do not need to measure distances and elevations along the entire transect. For example, for **Figure 20.3a**, you could just measure the distance and elevation from the 6400´ contour to the 6800´ contour.

1. Calculate the relief for each volcano by subtracting the lower elevation on the transect line from the higher elevation. Enter the relief into **Table 20.1**.

2. Calculate the length along each transect between the contour lines you used to measure the relief, using the scale bar. You can also calculate the length of each transect by measuring the transect with a ruler and using the written scale to convert it to feet. Enter the length into Table 20.1.

3. Calculate the slope in percentage by dividing the relief by the length. Enter the slope into Table 20.1.

TABLE 20.1 Volcano type

Figure	Relief	Length	Slope
A			
B			
C			

4. Compare the results you recorded in Table 20.1 with **Table 20.2**. The three volcanoes in this section are the Cinder Cone at Lassen Volcanic National Monument in California; Mauna Kea, a shield volcano on the Big Island of Hawai'i; and Mount Shasta, a composite volcano in California.

TABLE 20.2 Volcano type

Volcano Type	Slope and Relief	Tectonic Setting	Eruptive Behavior
Shield volcano	2°–10° (3%–18%), up to 10 km	Hot spot, rifting	Effusive
Cinder cone	34° (67%), up to 300 m	Hot spot, rifting	Lava fountain
Composite	Up to 30° (57%), over 4 km	Subduction	Explosive

121°19'30"W   121°19'0"W

6200

6900'
6650'
6400'
6150'
5900'

ELEVATION

A   B

40°33'0"N

6600

6400

B

40°33'0"N

6800

6400

6800

A

6600

6600

6400

600   500   400   300   200   100   0   600
FEET

300   200   100   0   300
METERS

SCALE 1: 5,000

121°19'0"W   40°32'30"N

121°19'30"W

▲ Figure 20.3a

▲Figure 20.3b

▲ Figure 20.3c

5. Which volcano is shown in Figure 20.3a? Explain your conclusion. Which tectonic setting is the source of its magma? What are the types of materials produced during eruptions and the hazards associated with these volcanoes?

6. Which volcano is shown in **Figure 20.3b**? Explain your conclusion. Which tectonic setting is the source of its magma? What are the types of materials produced during eruptions and the hazards associated with these volcanoes?

7. Which volcano is shown in **Figure 20.3c**? Explain your conclusion. Which tectonic setting is the source of its magma? What are the types of materials produced during eruptions and the hazards associated with these volcanoes?

8. Which volcano is composed of the most fluid lava? The least fluid lava? Explain.

9. In the boxes in Figures 20.5a, 20.5b, and 20.5c, draw an artistic rendering of a cross section of the tectonic situation that has created each of these volcanoes. For example, you could include a side view of tectonic rifting if one of the volcanoes had been made of flood basalt flows or a hot spot if one of the volcanoes were a shield volcano.

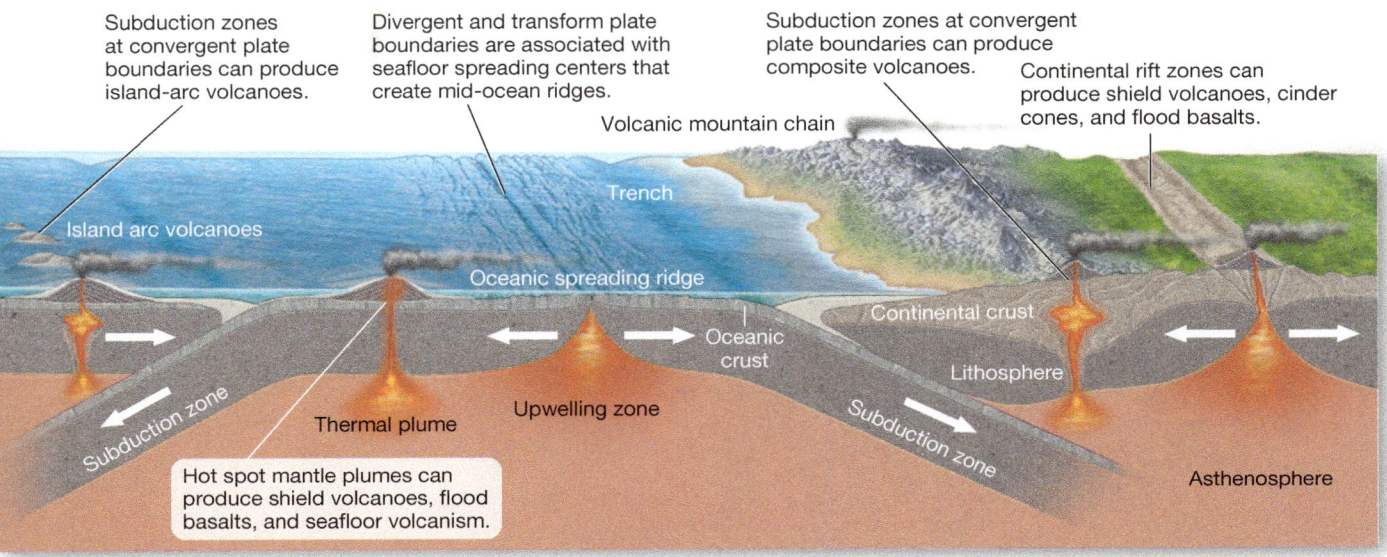

▲ **Figure 20.4**

▲ **Figure 20.5a**   Setting for volcano in Figure 20.3a

▲ **Figure 20.5b**   Setting for volcano in Figure 20.3b

▲ **Figure 20.5c** Setting for volcano in Figure 20.3c

SECTION 3

Kīlauea Caldera in Hawai'i

tual Tour
Kīlauea
Caldera

'/goo.gl/Amq4Ky

On the island of Hawai'i the continuing Kīlauea eruption represents the longest eruptive episode in Hawaiian history, active since 1983! During 1989–1990, lava flows from Kīlauea actually consumed several visitor buildings in the Hawai'i Volcanoes National Park and homes in Kalapana, a nearby town. Kīlauea has produced more lava than any other vent on Earth. Characteristic of a hot spot, eruptions emanating from the **asthenosphere** are low-viscosity magmas that are very fluid and yield a dark, basaltic rock (less than 50% silica and rich in iron and magnesium). Gases readily escape from this magma because of its texture, causing an effusive eruption that pours out on the surface, with relatively small explosions and little tephra. These are the relatively gentle eruptions that produce enormous volumes of lava annually on the sea floor and in places like Hawai'i. However, dramatic fountains of basaltic lava sometimes shoot upward, powered by jets of rapidly expanding gases.

An effusive eruption may come from a single vent or from a flank eruption through side vents in surrounding slopes. If such vents form a linear opening, they are called *fissures* and sometimes form a dramatic curtain of fire during eruptions. In Iceland, active fissures are spread throughout the plateau landscape. In Hawai'i, rift zones capable of erupting tend to converge on the central crater, or vent. The interior of such a crater is often a sunken caldera, which may fill with low-viscosity magma during an eruption, forming a molten lake, which may then overflow lava downslope to the sea.

Mauna Loa is an enormous shield volcano rising 5800 m (19,029 ft) from the sea floor to the surface and a further 4169 m (13,678 ft) above sea level—a total height of 9969 m (32,707 ft)! It is 18 km by 9 km (11 mi by 5.5 mi). **Figure 20.6** is a topographic map of a portion of the Kīlauea Volcano that has formed on the southeast slope of Mauna Loa. **Figure 20.7** is National Aerial Imagery Program (NAIP) imagery of the same area of the Kīlauea crater. **Figure 20.8** shows the dates of the most recent eruptions of Kīlauea crater.

Questions about and analysis of the Kīlauea Caldera.

▲ **Figure 20.6** Topographic Map of Kīlauea Crater, HI

155°17'30"   155°16'30"   155°15'30"

19°25'30"

0"

0"   19°24'30"

0"   19°23'30"

▲ **Figure 20.7** Aerial Imagery of Kīlauea Crater, HI

SCALE 1: 24,000

155°17'30"   155°16'30"   155°15'30"

1. What is the scale of this map?

   What is the contour interval?

2. What are the north-south length and east-west width of Kīlauea Crater?

   in miles:

   in kilometers:

3. What is the vertical relief (the difference between the highest and lowest elevations) between the volcano observatory at Uwekahuna Bluff on the rim of the crater and the floor of the crater at the 3524-foot bench mark? Note the tick marks on the contour lines.

4. Within the crater is a "fire pit" named Halemaumau. This pit has varied in size during eruptions. What is the relief from the rim of Halemaumau to the 3412-foot benchmark on the floor of the pit?

5. Record the dated lava flows noted on the map in the Kīlauea Crater, from oldest to youngest:

6. Describe how the appearance of the most recent lava flows compares to the older lava flows. What processes have caused this difference.

7. If you wanted to see the floor of Halema'uma'u when it was molten lava, when should you have been there?

8. What signs of recent volcanic activity do you see in Figure 20.7?

▲ Figure 20.8

▲ **Figure 20.9** Topographic map of Shiprock, New Mexico.

SHIP ROCK, NM

NEW MEXICO

QUADRANGLE LOCATION

CONTOUR INTERVAL 20 FEET

SCALE 1: 24,000

SECTION 4

Virtual Tour
Shiprock

https://goo.gl/3g7ED8

Shiprock Volcanic Neck

Shiprock is an interesting igneous rock formation in northwestern New Mexico. Magma (molten rock) flowed through a conduit to the surface, forming a volcanic cone. When eruptions ceased, the volcanic material eroded away, leaving the jagged lava plume standing starkly above the surrounding landscape—a volcanic neck of basalt. Also, magma flowed along linear cracks from the conduit, radiating outward to form **dikes**. Millions of years of erosion exposed this volcanic neck and radiating ridges composed of fine-grained basalt. To the Navajo Nation, the sacred 518 m (1700 ft) tall formation is the "rock with wings."

Using the topographic map in **Figure 20.9** and the aerial imagery in **Figure 20.10**, examine this volcanic feature. As you work with this image, complete the following questions.

1. Based only on the appearance of the surrounding landscape, does Shiprock appear to be composed of different rock or similar rock materials than surrounding rocks? Note the observations that led you to your opinion.

2. Is Shiprock an intrusive igneous feature or an extrusive igneous feature? Explain.

3. What is the relief from the top of Shiprock to the ground at the base of Shiprock?

4. What is the relief from the top of the west dike to the ground at the base of that dike? What is the relief from the highest part of the south dike to the ground at the base of that dike?

5. Describe the endogenic and exogenic processes, including physical and chemical weathering, that produced this volcanic formation.

6. Even though northwestern New Mexico is a desert, do you see any indication of water at work as an agent of erosion and transportation? Explain your observations from the aerial imagery.

Mastering Geography

Looking for additional lab prep and review materials? Visit the Study Area in Mastering Geography to enhance your geographic literacy and spatial reasoning skills by accessing MapMaster™ interactive maps, videos, *Mobile Field Trips*, *Project Condor* Quadcopter Videos, *In the News* articles, flashcards, web links, self-study quizzes, and other resources. Contact your professor to find out how you can access Mastering Geography.

▲ Figure 20.10 Aerial imagery of Shiprock, NM. Photos by NAIP, USGS, Eros Data Center.

SCALE 1: 24,000

Name: _____ Laboratory Section: _____

Date: _____ Score/Grade: _____

Video
Exercise 21
Pre-Lab Video

http://goo.gl/mQPkAt

Scan to view the
Pre-Lab video

LAB EXERCISE

The Rock Cycle and Rock Identification

A **rock** is usually an assemblage of **minerals** bound together (such as granite, a rock containing three minerals) but can also be a mass composed of a single mineral (such as rock salt), an undifferentiated material (such as the noncrystalline glassy obsidian), or even solid organic material (such as coal). A mineral is an inorganic natural solid compound having a specific chemical formula and usually a crystalline structure. Each mineral has its own characteristic color, texture, crystal shape, and density. There are more than 4200 known minerals, and of these about 30 are very common. Scientists have classified thousands of different rocks, all of which can be grouped into three types, based on the process that formed them.

Once we understand how a rock was formed, we gain insight into the environmental and tectonic settings present when it formed. **Igneous rocks** are rocks that solidified and crystallized from hot molten magma or lava, and we can tell whether they cooled quickly at the surface or slowly deep under the surface. Igneous rocks can also tell us if their magma was made of continental sediment or oceanic crust before melting. **Sedimentary rocks** are formed by the compaction, cementation, or hardening of sediments derived from other rocks; they tell us whether a region was a high-energy beach, a shallow inland sea, or a deep ocean trench. **Metamorphic rocks** are formed from existing igneous, sedimentary, or even metamorphic rock that has been transformed by going through profound chemical changes under pressure and increased temperature. Metamorphic rocks can tell us about the tectonic activity of a region, as well as about the conditions that formed the rock before it was transformed into a metamorphic rock. Lab Exercise 21 has two sections.

Key Terms and Concepts

extrusive igneous rock
felsic
igneous rock
intrusive igneous rock

mafic
metamorphic rock
mineral

rock
rock cycle
sedimentary rock

KEY**LEARNING**concepts

After completion of this lab, you should be able to:

- *Diagram* the rock cycle.
- *Describe* the three types of rock.
- *Identify* common rocks.

Materials/Sources Needed

magnifying glass

rock hand samples

Lab Exercise and Activities

SECTION 1

The Rock Cycle

The **rock cycle** is the name for the continuous alteration of Earth materials from one rock type to another (see **Figure 21.1**). For example, igneous rock formed from magma may break down into sediment by weathering and erosion and then lithify into sedimentary rock. This rock may subsequently become buried and exposed to pressure and heat deep within Earth, forming metamorphic rock. This may, in turn, break down and become a sedimentary rock again. Igneous rock may also take a shortcut through that cycle by directly becoming metamorphic rock. As the arrows in Figure 21.1 indicate, there are many pathways through the rock cycle. Any rock type can become any other rock type in only one step. Rocks can also cycle through the same process; for example, an igneous rock can melt and become a new igneous rock. Two cyclic systems drive the rock cycle. At and above Earth's surface, the hydrologic cycle, fueled by solar energy, drives the exogenic processes. Below Earth's surface and within the crust, the tectonic cycle, powered by internal heat, drives the endogenic processes.

(a) Quartz sand, Great Sand Dunes National Park, Colorado.

Compaction, cementation, chemical actions (lithification)

Sediment

Sedimentary rock

Weathering, erosion, transportation, deposition

Heat, intense pressure (metamorphism)

(b) Sedimentary layers in Kalamina Gorge, Western Australia.

Igneous rock

Metamorphic rock

Cooling, solidification (crystallization)

Melting

Magma

(d) An active lava flow, surrounded by basalt, Big Island, Hawai'i.

(c) Metamorphic quartzite on Catoctin Mountain near Thurmont, Maryland.

▲ **Figure 21.1** The rock cycle.
In this schematic of the relationships among igneous, sedimentary, and metamorphic processes, the thin arrows indicate "shortcuts"—such as when igneous rock is melted and becomes metamorphic rock without first going through a sedimentary stage. Note also that metamorphic rocks can undergo heating and pressure that causes further metamorphism without passing through the other stages. [(a) Marek Uliasz/Alamy. (b) blickwinkel/Alamy. (c) zrfphoto/iStock/Getty Images. (d) USGS.]

To identify a rock, one of the first steps is to determine which group it belongs to. Each rock has characteristics that allow it to be identified as igneous, sedimentary, or metamorphic. Relatively hard igneous rocks are generally made of randomly oriented, interlocking crystals of two or more minerals. Some have obvious physical characteristics, such as a dark, glassy appearance or many holes left by gas pockets that identify them as igneous rocks.

Sedimentary rocks are typically made of weathered material such as silt, sand, or pebbles. Some sedimentary rock is composed of fossils or plant debris, but other sedimentary rock is formed when minerals precipitate out of water, forming interlocking crystals. These rocks can be differentiated from crystalline igneous rocks because they are composed of readily identifiable minerals such as halite, calcite, or gypsum.

Metamorphic rocks are formed by the alteration of other rocks by heat and pressure. Some metamorphic rocks have a foliated appearance, caused by the parallel alignment of flat or elongated crystals, while others are made of stretched or deformed pebbles.

1. Place a check mark in the rock type column that corresponds to the process that could form or alter that rock.

TABLE 21.1 Rock types and processes			
Process	Igneous	Sedimentary	Metamorphic
Crystals left by evaporation of water			
Lithification of sediment			
Fossils present			
Melting of rock			
Folding of rock			
Foliated			
Heating but not melting			
Cooling of lava or magma			
Cementation of grains			
Compaction of sediment			

2. Identify a location in the United States where each step of the rock cycle is occurring. Describe what is happening at each location to create that rock type.

3. Describe a sequence in which a sedimentary rock forms and is modified to become a metamorphic rock.

SECTION 2

Rock Identification

To understand and classify rocks from a scientific viewpoint, we must begin with minerals, which are the building blocks of rocks. Identifying the minerals within a rock can help to identify that rock.

A rock is composed of a mineral or several minerals bound together, or solid organic material. The thousands of different rocks can be sorted according to three types, depending on the processes that formed them: igneous, sedimentary, and metamorphic.

Igneous Rocks

Approximately 90% of Earth's crust is made of igneous rocks, although sedimentary rocks, soil, or oceans frequently cover them.

Magma either intrudes into crustal rocks, cooling and hardening below the surface to form **intrusive igneous rock**, or it extrudes onto the surface as lava and cools to form **extrusive igneous rock**. The location and rate of cooling determine whether the rock is made of coarser (larger) or finer (smaller) crystals. Magma cools slowly below the surface, which allows more time for crystals to grow, resulting in coarse-grained rocks such as granite, which can be exposed to the surface by later tectonic uplift. Lava cools quickly at the surface and forms finer-grained rocks, such as basalt, the most common extrusive igneous rock. Basalt makes up 71% of the ocean floor, and it is actively forming the Big Island of Hawaii. If cooling is so rapid that crystals cannot form, the result is a glassy rock such as obsidian, or volcanic glass. Pumice does not have a crystal structure, and it forms when the bubbles from escaping gases create a frothy texture in the lava. Pumice is full of small holes, is light in weight, and is low enough in density to float in water.

Igneous rocks are classified according to their texture and composition. The same magma that produces coarse-grained granite when it cools beneath the surface can form fine-grained rhyolite when it cools above the surface. The mineral composition of a rock, especially the relative amount of silica (SiO_2), provides information about the source of the magma that formed it and affects its physical characteristics. **Felsic** igneous rocks, such as granite, are high in silicate minerals, such as *fel*dspar and quartz (pure *si*lica). Rocks formed from felsic minerals generally are lighter in color, less dense than those from mafic minerals, have lower melting points, and are less fluid when molten.

Mafic igneous rocks, such as basalt, are derived from *ma*gnesium and iron (the Latin word for iron is *fer*rum). Compared to felsic magma, mafic rocks are lower in silica, higher in magnesium and iron, have high melting points, and are much more fluid when molten. Rocks formed from mafic minerals are darker in color and of greater density than those from felsic minerals. Ultramafic rocks have the lowest silica content; examples are peridotite (less than 45% silica) and the rocks of the mantle.

Condor
Intrusive
Igneous Bodies

https://goo.gl/IdkeDC

Sedimentary Rocks

Sedimentary rock can be formed from loose grains or fragments called *clasts* that are cemented together. Clasts come from several sources: the weathering and erosion of existing rock (the origin of the sand that forms sandstone), shells on the ocean floor (which make up one form of limestone), organic matter from ancient plants (which form coal), and the precipitation of minerals from water solution (the origin of the calcium carbonate, $CaCO_3$, that forms chemical limestone, although most limestone is biogenic). Sedimentary rocks are divided into several categories—clastic, biochemical, organic, and chemical—based on their origin.

Clastic Sedimentary Rocks

The formation of clastic sedimentary rock involves several processes. Weathering and erosion disintegrate and dissolve existing rock into clasts, which are carried across the landscape by gravity, water, wind, and ice. The clasts are now called *sediment*. Transport occurs from "higher-energy" sites, where the carrying medium has the energy to pick up and move the material, to "lower-energy" sites, where the sediment is deposited. Deposition is the process whereby sediment settles out of the transporting medium and results in material dropped along river channels, on beaches, and on ocean bottoms, where it is eventually buried. Lithification occurs as loose sediment is hardened into solid rock. This process involves compaction of buried sediments as the weight of overlying material squeezes out the water and air between clasts and cementation by minerals, which fill any remaining spaces and fuse the clasts—principally quartz, feldspar, and clay minerals—into a coherent mass. The type of cement varies in different environments. Calcium carbonate ($CaCO_3$) is the most common cement, followed by iron oxides and silica. Drying and heating can also unite particles. The different sediments that make up sedimentary rock range in size from boulders to gravel to sand to microscopic clay particles. After lithification, these size classes, combined with their composition, sorting, and cement characteristics, determine the common sedimentary rock types. For example, pebbles and gravels become conglomerate, silt-sized particles become siltstone or mudstone, and clay-sized particles become shale.

Biochemical, Organic, and Chemical Sedimentary Rocks

Some sedimentary rocks are formed from calcium carbonate from the shells of organisms (a biochemical process) or from dissolved minerals that precipitate out of water (a chemical process) and build up to form rock. Chemical precipitation is the formation of a separate solid substance from a solution, such as when water evaporates and leaves behind a residue of salts. The most common chemical sedimentary rock is limestone, and the most common form of limestone is biochemical limestone from marine organic origins. When calcium carbonate shelled organisms die, the solid material builds up on the ocean floor and lithifies to become limestone. The salt deposits left by evaporating water can build up to form evaporites, a type of chemical sedimentary rock. Examples of these evaporites are found in Utah on the Bonneville Salt Flats, created when an ancient lake evaporated. Both clastic and chemical sedimentary rocks are deposited in layered strata that form an important record of past ages. Using the principle of superposition, scientists use the stratigraphy (the ordering of layers), thickness, and spatial distribution of strata to determine the relative age and origin of the rocks. The strata also indicate the region's climatic history, since each layer was formed under different environmental conditions.

Metamorphic Rocks

Igneous, sedimentary, or even metamorphic rock may be transformed into a metamorphic rock by going through profound physical or chemical changes under pressure and increased temperature. The name metamorphic comes from a Greek word meaning "to change form." Metamorphic rocks generally are more compact than the original rock and therefore are harder and more resistant to weathering and erosion. Some of the processes that can cause metamorphism are heating, pressure, heating and pressure together, and hydrothermal metamorphism—such as along subduction zones and mid-ocean ridges. When heat is applied to rock, the atoms within the minerals may break their chemical bonds, move, and form new bonds, leading to new groupings of minerals. When pressure is applied to rock, its structure may change as atoms become packed more closely. When rock is subject to both heat and pressure at depth, the original mineral assemblage becomes unstable and changes. Finally, rocks may be compressed by overlying weight and subject to shear when one part of the mass moves sideways relative to another part. These processes change the shape of the rock, leading to changes in the mineral alignments within. Metamorphic rock may form from igneous rocks as the lithospheric plates shift, especially when one plate is thrust beneath another. Contact metamorphism occurs when magma rising within the crust "cooks" adjacent rock; this type of metamorphism results from heat alone. Regional metamorphism occurs when a large areal extent of rock is subject to metamorphism. This can occur when sediments collect in broad depressions if their own weight creates enough pressure in the bottommost layers to transform the sediments into metamorphic rock. Regional metamorphism also occurs as lithospheric plates collide and mountain building occurs. Metamorphic rocks have textures that are foliated or nonfoliated, depending on the arrangement of minerals after metamorphism. Foliated rock has a banded or layered appearance, demonstrating the alignment of minerals, which may appear as wavy striations (streaks or lines) in the new rock. Nonfoliated rocks do not exhibit this alignment.

1. What characteristics indicate whether a rock is an intrusive or extrusive igneous rock?

2. What characteristics would indicate whether a rock is a clastic or nonclastic sedimentary rock?

3. Explain how a metamorphic rock could develop a foliated appearance or a nonfoliated appearance.

Use the following rock identification keys; **Figures 21.2, 21.3** and **21.4**, and the images of rock samples in **Figure 21.5** to identify the rock samples that your instructor provides. Fill out **Tables 21.2, 21.3**, and **21.4** as you analyze the rocks.

IGNEOUS ROCK IDENTIFICATION KEY

		MINERAL COMPOSITION			
		Felsic (Granitic)	**Intermediate** (Andesitic)	**Mafic** (Basaltic)	**Ultramafic**
Dominant Minerals		Quartz Potassium feldspar	Amphibole Plagioclase feldspar	Pyroxene Plagioclase feldspar	Olivine Pyroxene
Accessory Minerals		Plagioclase feldspar Amphibole Muscovite Biotite	Pyroxene Biotite	Amphibole Olivine	Plagioclase feldspar
TEXTURE	**Coarse-grained**	Granite	Diorite	Gabbro	Peridotite
	Fine-grained	Rhyolite	Andesite	Basalt	Komatiite (rare)
	Porphyritic (two distinct grain sizes)	Granite porphyry	Andesite porphyry	Basalt porphyry	Uncommon
	Glassy	Obsidian	Less common	Less common	Uncommon
	Vesicular (contains voids)	Pumice (also glassy)		Scoria	Uncommon
	Pyroclastic (fragmental)	Most fragments < 4mm → Tuff or welded tuff	← →	Most fragments > 4mm Volcanic breccia	Uncommon
Rock Color (based on % of dark minerals)		0% to 25%	25% to 45%	45% to 85%	85% to 100%

▲ Figure 21.2 [Images: Dennis Tasa or Edward J. Tarbuck.]

SEDIMENTARY ROCK IDENTIFICATION KEY

Detrital Sedimentary Rocks

Clastic Texture (particle size)		Distinctive Properties	Rock Name	
Gravel Coarse (over 2 mm)		Rounded rock or mineral fragments, typically poorly sorted	Conglomerate	
		Angular rock or mineral fragments, typically poorly sorted	Breccia	
Sand Medium (1/16 to 2 mm)		Quartz grains, typically rounded, well sorted	Quartz Sandstone	Sandstone
		At least 25% feldspar, typically poorly sorted, angular fragments	Arkose	
		Mixture of sand and mud, typically poorly sorted	Graywacke	
Silt/mud Fine (1/16 to 1/256 mm)		Mostly silt-size quartz and clay, blocky, gritty	Siltstone	Mudstone
Clay/mud Very fine (less than 1/256 mm)		Mostly clay, splits into layers, may contain fossils	Shale	
		Mostly clay, crumbles easily	Claystone	

Chemical Sedimentary Rocks

Composition	Distinctive Properties	Rock Name	
Calcite, $CaCO_3$ (effervesces in HCl)	Fine to coarse crystalline	Crystalline Limestone	Limestone
	No visible grains, may exhibit conchoidal fracture	Micrite	
	Visible shells and shell fragments loosely cemented	Coquina	
	Various size shells and shell fragments, well cemented	Fossiliferous Limestone	
	Microscopic shells and clay, soft	Chalk	
	Faint layering, may contain cavities or pores	Travertine	
Quartz, SiO_2	Microcrystalline, may exhibit conchoidal fracture, will scratch glass	Chert (light colored)	
		Flint (dark colored)	
		Agate (banded)	
Gypsum $CaSO_4 \cdot 2H_2O$	Fine to coarse crystalline, soft	Rock Gypsum	
Halite, NaCl	Fine to coarse crystalline, tastes salty	Rock Salt	
Altered plant fragments	Black brittle organic rock, may be layered	Bituminous Coal	

▲ Figure 21.3

METAMORPHIC ROCK IDENTIFICATION KEY

Distinctive Properties	Grain Size	Parent Rock	Texture	Rock Name
Excellent rock cleavage, smooth dull surfaces	Very fine	Shale, or siltstone		Slate
Breaks along wavy surfaces, glossy sheen	Fine	Shale, slate, or siltstone		Phyllite
Micas dominate, breaks along scaly foliation	Medium to Coarse	Shale, slate, phyllite, or siltstone		Schist
Compositional banding due to segregation of dark and light minerals	Medium to Coarse	Shale, schist, granite, or volcanic rocks		Gneiss
Interlocking calcite or dolomite crystals nearly the same size, soft, reacts to HCl	Medium to coarse	Limestone, dolostone		Marble
Fused quartz grains, massive, very hard	Medium to coarse	Quartz sandstone		Quartzite
Round or stretched pebbles that have a preferred orientation	Coarse-grained	Quartz-rich conglomerate		Metaconglomerate
Shiny black rock that may exhibit conchoidal fracture	Fine	Bituminous coal		Anthracite
Usually, dark massive rock with dull luster	Fine	Any rock type		Hornfels
Very fine grained, typically dull with a greenish color, may contain asbestos fibers	Fine	Mafic or ultramafic rocks		Serpentinite

▲ Figure 21.4 [Images: Dennis Tasa or Edward J. Tarbuck.]

TABLE 21.2 Igneous rocks

IGNEOUS ROCKS

Sample Number	Texture	Color	Mafic or Felsic	Rock Name
1				
2				
3				
4				
5				
6				
7				
8				
9				
10				

TABLE 21.3 Sedimentary rocks

SEDIMENTARY ROCKS

Sample Number	Texture or Grain Size	Composition or Sediment Name	Detrital or Chemical (Biochemical)	Rock Name
1				
2				
3				
4				
5				
6				
7				
8				
9				
10				

TABLE 21.4 Metamorphic rocks

METAMORPHIC ROCKS

Sample Number	Grain Size	Foliated or Nonfoliated	Rock Name	Parent Rock
1				
2				
3				
4				
5				
6				
7				
8				
9				
10				

Mastering Geography

Looking for additional lab prep and review materials? Visit the Study Area in Mastering Geography to enhance your geographic literacy and spatial reasoning skills by accessing MapMaster™ interactive maps, videos, *Mobile Field Trips, Project Condor* Quadcopter Videos, *In the News* articles, flashcards, web links, self-study quizzes, and other resources. Contact your professor to find out how you can access Mastering Geography.

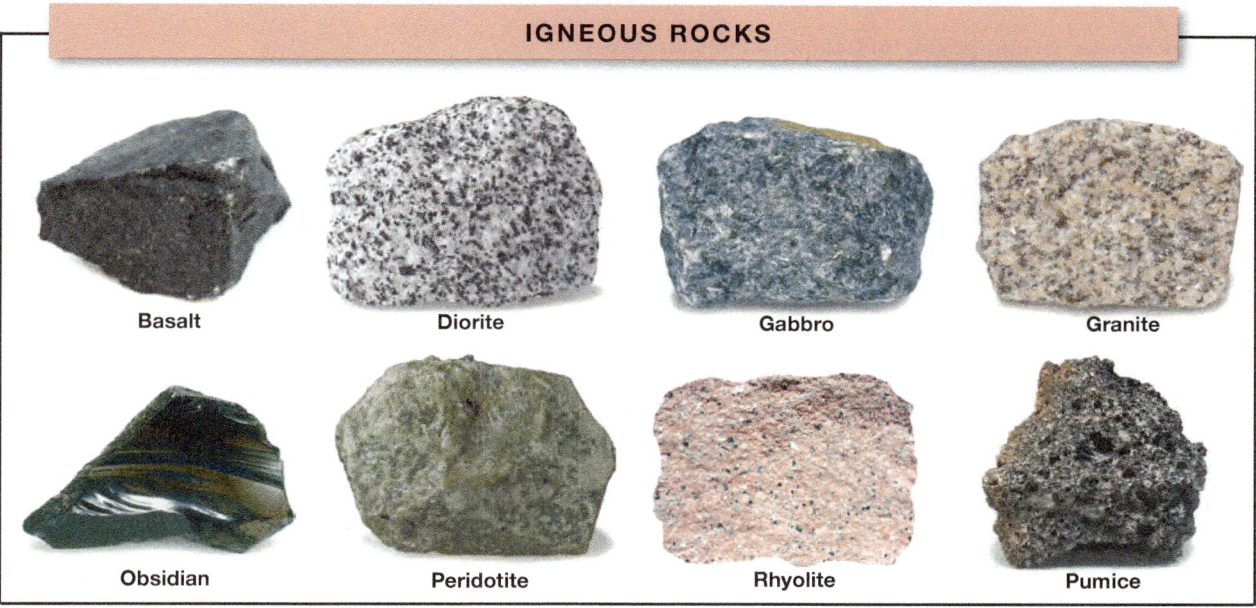

IGNEOUS ROCKS

Basalt

Diorite

Gabbro

Granite

Obsidian

Peridotite

Rhyolite

Pumice

SEDIMENTARY ROCKS

Breccia

Chert

Conglomerate

Limestone

Sandstone

Shale or Mudstone

Fossiliferous limestone

METAMORPHIC ROCKS

Gneiss

Marble

Quartzite

Schist

Serpentinite

Slate

Stretched pebble conglomerate

Phyllite

▲ **Figure 21.5** Rock Samples [Basalt: Tyler Boyes/Shutterstock; Phyllite: USGS; Serpentinite: Charlie Thomsen; All others: Dennis Tasa or Edward J. Tarbuck.]

NOTES

Name: _____ Laboratory Section: _____

Date: _____ Score/Grade: _____

LAB EXERCISE

22

Recurrence Intervals for Natural Events

Geographers compile spatial data in order to recognize patterns among natural events, make predictions, and encourage preventive or protective actions in an attempt to mitigate the impact of various environmental hazards. Using statistical data to determine the probability of recurrence of a natural event is the focus of this lab exercise. Probabilities are important in guiding protective policies, planning, and hazard zoning, if allowed by the usually sluggish political process. In this lab exercise, we focus on stream processes and the recurring floods that endanger society.

Hydrology is the science of water, its global circulation, distribution, and properties, specifically water at and below Earth's surface. For hydrology online info, see http://nwis.waterdata.usgs.gov/nwis or http://water.usgs.gov. A **recurrence interval**, or *return period*, is the average time period within which a given event will be equaled or exceeded once. This exercise examines annual streamflow data to establish recurrence intervals for flooding and to see how this can be used for improved planning and disaster management. Lab Exercise 22 has three sections.

Key Terms and Concepts

discharge
flood
flood frequency curve
flood probability

gage height (stage)
hydrology
rating curve

recurrence interval (return
   period)
streamflow

KEY**LEARNING**concepts

After completion of this lab, you should be able to:

- **Define** stream discharge and gage height and *describe* the relationship between them.
- **Use** selected hydrographic data to construct rating curves to *illustrate* streamflow characteristics.
- **Define** recurrence intervals and use hydrographic data to *calculate* and *graph* them.
- **Use** rating curves and flood frequency curves to *assess* risk potential and *make* land-use decisions based on flood potential.

Materials/Sources Needed

pencils
calculator

colored pencils
ruler

Lab Exercise and Activities

SECTION 1

Streamflow, Discharge, and Gage Height

Stream channels vary in *width* and *depth*. The streams that flow in them vary in *velocity* and in the *sediment load* they carry. All these factors increase with increasing **discharge**, or the stream's volume of flow per unit time. Discharge is calculated by multiplying the velocity of a stream by its width and depth for a specific cross section of the channel, as stated in the simple expression:

$$Q = wdv$$

where Q = discharge, w = channel width, d = channel depth (measured above an arbitrary datum called the stage), and v = steam velocity. As Q increases, some combination of channel width, depth, and stream velocity increases. Discharge is expressed in cubic meters per second (cms, or m³/sec) or cubic feet per second (cfs, or ft³/sec). Given the interplay of channel width and depth and

stream velocity, the cross section of a stream varies over time, especially during heavy floods. Flood avoidance or management requires extensive measurement of streamflow characteristics and discharge (flow per unit of time).

Table 22.1 contains data from 1882 to 2016 for the Red River of the North, collected at the gaging station at Grand Forks, North Dakota. The maximum peak discharge and corresponding **gage height** in meters are shown for the selected years. **Stage** is also called stream stage or gage height and is the height of the water surface, in meters, above an established elevation where the stage is zero. This zero elevation is based on a reference elevation datum, either the National Geodetic Vertical Datum of 1929 (NGVD29) or the later North American Vertical Datum of 1988 (NAVD88).

Notes for Table 22.1:
Station name: Red River of the North at Grand Forks, Grand Forks County, ND
Station number: 05082500
Latitude: 47.93°N
Longitude: 97.03°W
Basin name: Sandhill-Wilson

Drainage area (square kilometers): 77,959 km²

Contributing drainage area (square kilometers): 68,117 km²
Gage datum elevation: 126 m (above NGVD 29)
Base discharge (cms): 237 cms

Gage heights are given in meters above gage datum elevation.
Discharge is listed in the table in cubic meters per second.
Peak flow data were retrieved from the USGS National Water Information System (NWIS).

Using the data provided in Table 22.1, answer the following questions and completion items.

1. **a)** The highest discharge recorded for the listed years: _____

   b) Year recorded: _____

   c) The lowest discharge recorded for the listed years: _____

   d) Year recorded: _____

2. **a)** Average discharge (1984–2013): _____

   b) Range of discharge (1984–2013): _____

TABLE 22.1 Peak annual discharge (in cubic meters per second) and gage height (in meters above gage datum elevation) for the Red River of the North at Grand Forks, North Dakota, 1882–2016. (USGS National Water Information Service, NWIS, http://nwis.waterdata.usgs.gov/usa/nwis/peak)

1	2	3	4	5
Peak Discharge Date	Discharge (cms)	Gage at Peak	Rank	Recurrence Interval
April 18, 1997	3836	15.9	1	135.00
April 14, 2011	2450	15.2	2	
April 10, 1897	2380	15.3	3	
April 23, 1979	2296	14.8	4	
April 1, 2009	2148	15.0	5	
April 18, 1882	2100	14.6	6	
April 6, 2006	2038	14.6	7	
March 20, 2010	1761	14.0	8	
April 21, 1996	1635	14.0	9	
April 14, 2001	1618	13.7	10	
April 13, 1989	1109	13.2	20	6.8
April 27, 1904	924	12.4	30	4.5
April 12, 1943	790	11.6	40	3.4
April 3, 1976	661	10.5	50	2.7
April 6, 1917	554	9.9	60	2.3
March 24, 1929	479	8.6	70	1.9
July 2, 1957	412	7.5	80	1.7
April 2, 1928	342	6.6	90	1.5
June 12, 1925	271	5.8	100	1.4
March 28, 1926	216	5.5	110	1.2
April 8, 1912	132	3.9	120	1.1
April 1, 1889	84	3.7	130	1.0
April 10, 1931	46	2.0	135	1.0

3. a) The highest gage at peak discharge: _____

   b) Year recorded: _____

   c) The lowest gage at peak discharge: _____

   d) Year recorded: _____

4. Scan the *Discharge* (column 2) and *Gage at Peak* (column 3) data columns. As gage height at peak (stream depth) gets higher, what usually happens to the discharge?

5. In general, there is a direct relationship between discharge and width, depth (gage height), and velocity. A stream's **rating curve** relates stream discharge to the gage height. **Figure 22.1** shows the gage height plotted against discharge. Note that the discharge axis is logarithmic.

   a) We can use the graph in Figure 22.1 to read data from the chart. Use the **rating curve** to determine the gage height associated with the following discharges. (*Hint:* Use colored pencils to mark them on the graph first.)

   100 cms _____[3.0 m]_____

   500 cms _____

   1000 cms _____

   3000 cms _____

   b) Determine the discharge associated with a gage height of

   6 meters _____[225 cms]_____

   9 meters _____

   12 meters _____

   15 meters _____

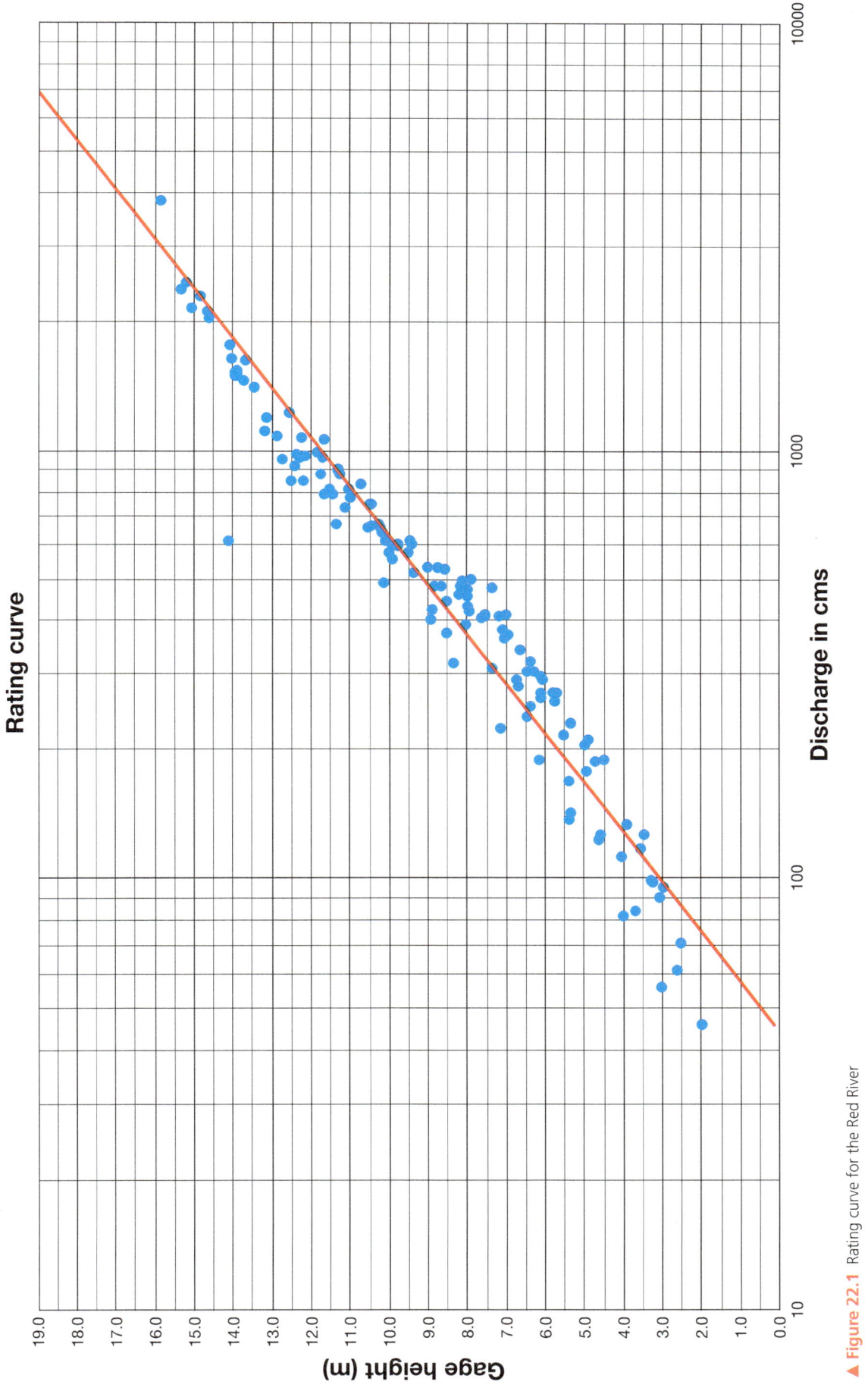

Rating curve

Gage height (m)

Discharge in cms

▲ Figure 22.1 Rating curve for the Red River

SECTION 2

Floods and Recurrence Intervals

Flood patterns in a drainage basin are as complex as the weather, for floods and weather are equally variable, and both include a level of unpredictability. Measuring and analyzing the behavior of each large watershed and stream enables engineers and concerned parties to develop the best possible flood-management strategy. The key to flood avoidance or management is to possess extensive measurements of **streamflow**, a stream's discharge, and how it performs during a precipitation event.

A **flood** is a high water level that overflows the natural river bank along any portion of a stream. Both floods and the floodplains they might occupy are rated statistically for the expected time intervals between floods.

A recurrence interval, or return period, is the average time period within which a given event will be equaled or exceeded once. A 10-year flood is the highest flood level (gage height) that is likely to occur once every 10 years; a 50-year flood is the highest that is likely to occur once every 50 years. A 10-year flood has a 10% likelihood of being equaled or exceeded in any one year and is likely to occur about 10 times in each century. It is important to note that these probabilities estimate *random recurrence*. Several decades may pass without a 10-year flood, or floods of such magnitude may occur two or three years in a row. The higher the return interval, the more catastrophic the event is likely to be and the less likely the probability of the event occurring in any given year.

Initial reports on the Midwest floods of 1993 and the Red River floods of 1997 included various references to 300-year and 500-year floods. Data now show that the 1993 Midwest floods easily exceeded a 1000-year **flood probability**. Let us now calculate the recurrence interval for the 1997 Red River floods.

1. Refer once again to Table 22.1 and at the far right locate columns 4 and 5, *Rank* and *Recurrence Interval*, which have not been used as yet. Rank the floods (using the gage height data in column 3), with the rank of 1 for the largest flood and continue in descending order to the smallest, entering your ranking numbers in column 5.

The recurrence interval (*RI*) is calculated using the following formula:

$$RI = \frac{n + 1}{m}$$

Where

$$RI = \text{recurrence interval (in years)}$$
$$n = \text{number of years of record}$$
$$m = \text{rank of the flood}$$

For example, the flood with a rank of 1 and 30 years of records would have a recurrence interval calculated as follows:

$$RI = \frac{30 + 1}{1} = 31$$

2. Calculate the rank and recurrence intervals for the gage heights and enter your results in column 5 of Table 22.1. Some (namely, flood ranks 1 and 10–116) have been done for you on the table to get started.

3. On the logarithmic paper in **Figure 22.2**, plot the recurrence (return) interval against the discharge. The sample points for flood ranks 1 and 11–116 have been plotted.

Flood frequency curve

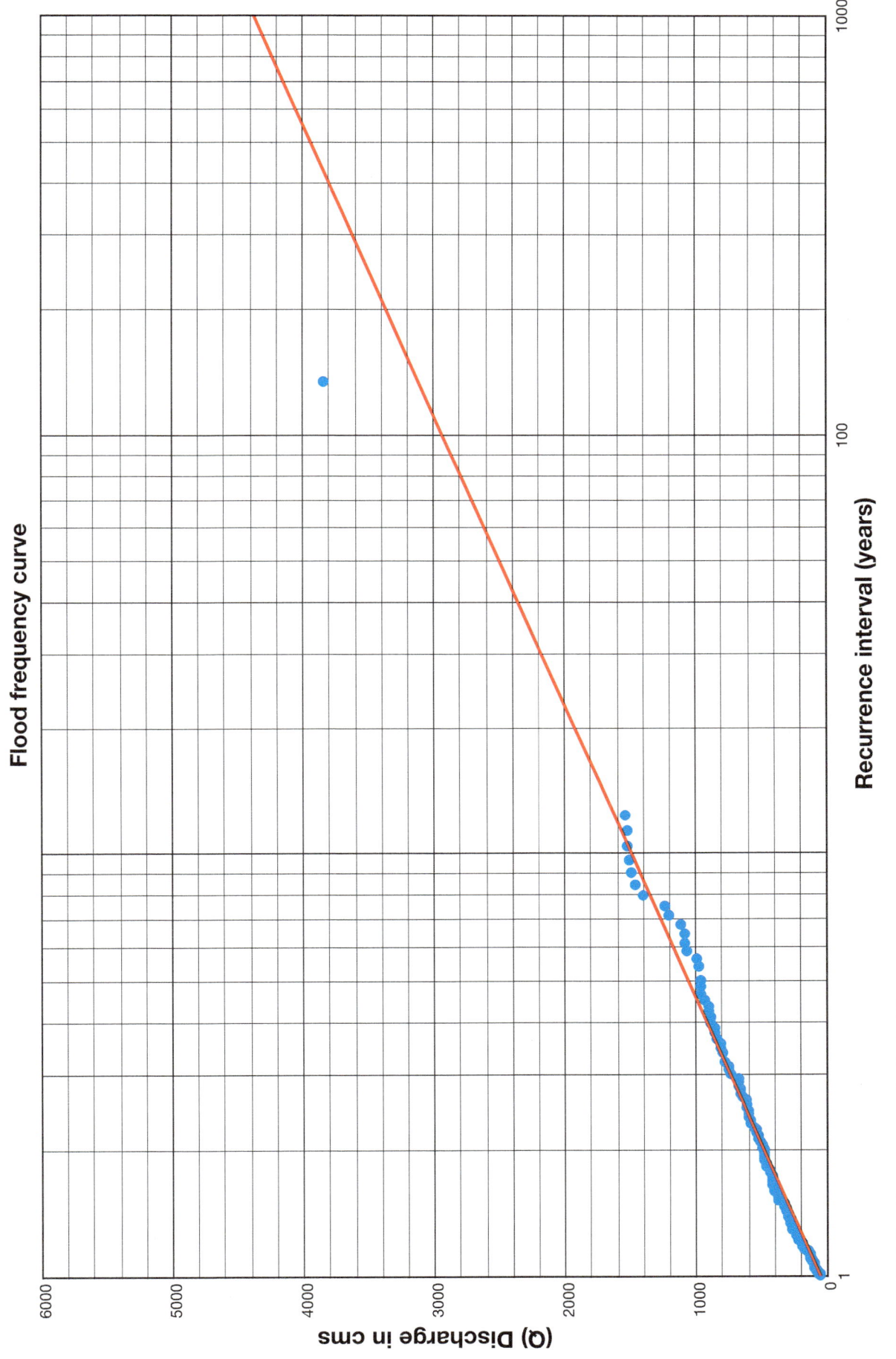

Discharge (Q) in cms *(vertical axis: 0, 1000, 2000, 3000, 4000, 5000, 6000)*

Recurrence interval (years) *(horizontal axis: 1, 100, 1000)*

Once again you can use the graphs you created to obtain information about hypothetical discharges and recurrence intervals.

4. **a)** Using the **flood frequency curve** in Figure 22.2 (and colored pencils to mark on graph), estimate the recurrence interval of a flood of the following magnitudes:

200 cms _____

1500 cms _____[10 years]_____

3000 cms _____

b) Estimate the discharge of the

50-year flood _____

100-year flood _____

500-year flood _____

5. **a)** Use Figure 22.2 along with Figure 22.1 to determine the gage height of these flood frequencies. (Use the flood frequency curve in Figure 22.2 to obtain the discharge, and then use that discharge on the rating curve to determine the associated gage height.)

20-year flood _____

300-year flood _____

400-year flood _____

b) Now use both graphs (Figure 22.1 first, and then 22.2) to determine the recurrence interval of the gage height at

10 meters _____

12 meters _____

15 meters _____

6. How many times in a century would you expect the Red River to crest at a gage height of 13 meters?

What is the probability of a flood of that magnitude in any given year?

7. According to the flood frequency curve, what is the return interval for a flood the magnitude of the spring 1997 flood?

SECTION 3

Probability, Prediction, and Risk Assessment

Throughout history, civilizations have settled floodplains and deltas, especially since the agricultural revolution of 10,000 years ago, when the fertility of floodplain soils was discovered. We build artificial levees to keep rivers within their channels and construct dams to hold back and regulate streamflow. How high must the levees and dams be? What type of zoning of the floodplain is required to safeguard settlements? Hydrographic data are compiled and the gage height and flood frequency curves are plotted in order to determine flood risk potential.

Red River, Grand Forks

The levees along the Red River at Grand Forks were 15.5 m (50.8 ft) high. In one 1997 event the gage height at peak discharge measured 15.86 m (52 ft), resulting from unusual icing conditions upstream, similar to a "dam break." Four days later the river rose even higher, to its peak gage height for the year, measuring 16.56 m (54.4 ft). Within a few days the river topped the levees twice, causing over $1 billion in damage!

Engineers have raised the Grand Forks levees to 18.2 m (60 ft), including an added lift of 0.91 m (3 ft) to allow for wave action at times of flooding. This was a $403 million project, shared by the citizens of Grand Forks. Ironically, other cities at similar risk along the river have not been as farsighted. This uneven mitigation effort leads to friction and problems between cities and governments when flood events occur.

Rivers have been important in the geography of human history, influencing where settlements were built, where livelihoods were made, and where borders were drawn. But to what extent should the general public continue to subsidize these risky sites? Rating curves and flood frequency curves can help us make wiser decisions about floodplain settlement and hazard zoning specifications.

After reading the above introduction, refer to question 4 in Section 2.

1. What is the recurrence interval for a flood the magnitude of the 1997 flood, based on your plotted line? __[425 years]__

2. According to your charts, what is the recurrence interval for Grand Forks' peak gage height for that year?

3. What are the probabilities of floods of these magnitudes occurring in any given year?

   April 18, 1997 (15.86 m) _____

   April 22, 1997 (16.56 m) _____

4. What would be the recurrence interval for a flood to reach the top of the new levee? _____

5. How high above the gage datum would you want to locate a new house in Grand Forks to make certain it would be safe from a 30-year flood? _____

Mastering Geography

Looking for additional lab prep and review materials? Visit the Study Area in Mastering Geography to enhance your geographic literacy and spatial reasoning skills by accessing MapMaster™ interactive maps, videos, *Mobile Field Trips, Project Condor* Quadcopter Videos, *In the News* articles, flashcards, web links, self-study quizzes, and other resources. Contact your professor to find out how you can access Mastering Geography.

NOTES

Video
Exercise 23
Pre-Lab Video

http://goo.gl/Y7mNBn

Scan to view the
Pre-Lab video

LAB EXERCISE

Contours and Topographic Maps

23

The character and general configuration of Earth's surface is called **topography**. An important diagnostic tool for landform analysis is the **topographic map**, a map that shows physical landscape features and elevation. While specific application and analysis of topographic maps relating to various types of landforms will be dealt with in later exercises, you have already explored the basic concepts of topographic maps when you analyzed temperature and pressure patterns. **Isolines**, lines connecting points of equal value, are seen on topographic maps as **contour lines**, lines connecting points of equal elevation. These isolines are conceptually similar to the isotherms (equal temperatures) and isobars (equal atmospheric pressure) that you used in Lab Exercises 9 and 11, as well as isohyets (equal precipitation). Therefore, a familiarity with reading and interpreting isolines will be useful at this point. A general discussion of topographic maps and contour lines is the focus of Lab Exercise 23. Lab Exercise 23 features mobile-friendly QR links to Google Earth virtual tours that fly you to the topographic map location, as well as optional online Google Earth activities with downloadable media and worksheets at mygeoscienceplace.com and the Mastering Geography Study Area. These optional questions allow you to compare the topographic maps and your work in the lab manual with Google Earth™ imagery and topographic maps draped over the landscape.

Key Terms and Concepts

Alber's projection
contour interval
contour lines
geographic index number
index contours
isogonic map
isolines

local relief
map view
planimetric map
profile view
relief
topography

topographic map
topographic profile
Universal Transverse Mercator (UTM) grid
vertical datum
vertical exaggeration

KEY**LEARNING**concepts

After completion of this lab, you should be able to:

- **Construct** contour lines (isolines of equal elevation) and *interpret* a mapped landscape.
- **Construct** a topographic profile and *calculate* an appropriate vertical exaggeration for a profile.
- **Describe** and *use* the legend and marginal labels and information on a topographic quadrangle map.

Materials/Sources Needed

pencils
colored pencils
calculator
Google Earth

compass
protractor
ruler

Lab Exercise and Activities

SECTION 1

Contour Lines and Topographic Maps

The USGS depicts information on *quadrangle maps*, rectangular maps bounded by parallels and meridians rather than by political boundaries. A conic map projection, the **Alber's projection** (equal-area), is used as a base for these quadrangle maps (see outside of fold-out flap on back cover of this manual). Two standard parallels (where the cone intersects the globe's surface) are used to improve the accuracy in conformality and scale for the conterminous United States: 29.5°N and 45.5°N latitudes (**Figure 23.1**). For the U.S. topographic mapping program, the standard parallels are shifted for conic projection base maps of Alaska (55°N and 65°N) and Hawai'i (8°N and 18°N).

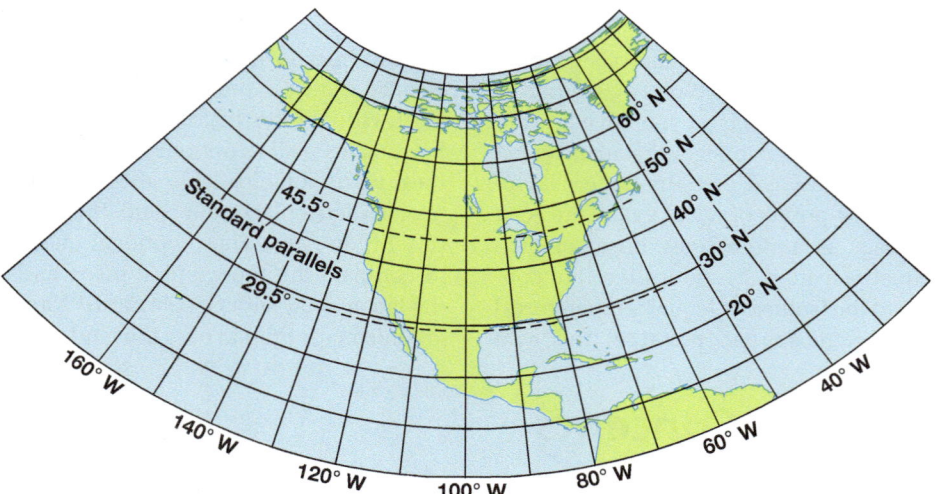

▲ **Figure 23.1** Alber's equal-area conic projection with standard parallels of 29.5°N and 45.5°N, a standard for U.S. topographic maps

In mapping, a basic **planimetric map** is first prepared, showing the horizontal position of boundaries; land-use aspects; and political, economic, and social features. A highway map, such as one that you may have in your car, is a common example of a planimetric map. A vertical scale of physical features is then added to portray the terrain. The most popular and widely used of these maps are the detailed topographic maps prepared by the USGS.

Topographic maps portray physical **relief**, the difference in elevation between the highest point and lowest point on the map. **Local relief** refers to the change in elevation between two specified points on the map. Relief is indicated through the use of elevation contour lines. A contour line connects all points at the same elevation above or below a stated reference level. This reference level—usually mean sea level—is called the **vertical datum**. The **contour interval** is the difference in elevation between two adjacent contour lines. (In **Figure 23.2**, the contour interval is 20 ft or 6.1 m; in **Figure 23.3**, the contour interval is 40 ft or 12.2 m.)

The topographic map in Figure 23.2 shows a hypothetical landscape that demonstrates how contour lines and intervals depict slope and relief. Slope is indicated by the pattern of lines and the space between them. The steeper a slope or cliff, the closer together the contour lines appear in **Figure 23.2b**; note the narrowly spaced contours representing the cliff. A more gradual slope is portrayed by a wider spacing of these contour lines, as you can see from the widely spaced lines on the beach. Once you become adept at reading contour lines, you will be able to "see" the relief, even though the maps are only in two dimensions.

Contour Lines

a. Each contour line connects points of equal elevation.
b. Contour lines are typically printed in brown color on topographic maps.
c. Every fifth contour is an index contour, with a thicker line width and its elevation labeled.
d. Contour lines separate areas of higher (upslope) elevation from areas of lower (downslope) elevation.
e. Varying degrees of steepness are obvious: The steepest areas are where the lines are closest together, and widely spaced contours depict where the slope is gentle.
f. Contour lines crossing stream valleys form a V pointing upstream.
g. Hills are indicated by a series of closed, concentric contours.
h. Given a large enough map, contour lines always form a closed polygon, although on a smaller map contour lines may run off the margin.
i. The contour interval is the difference in elevation between two adjacent contour lines.
j. Contour lines only touch where there is a steep cliff and never cross unless the cliff face has an overhanging ledge. (Hidden contours are then depicted as dashed lines.)
k. Concentric, closed contours denote a hill or summit, and a similar pattern with hachure marks (small tick marks) on the downslope side depicts a closed depression.
l. When a map is showing a depression, the first hachured contour line is the same elevation as the adjacent lower contour line.
m. Contour lines occur in pairs from one side of a valley to the other side.
n. Index contours are thicker and usually have their elevation labeled.

Use the hypothetical landscape and topographic map in Figure 23.2 to answer the following questions:

1. What is the contour interval?

   How can you tell?

2. In terms of local relief (the difference between the highest and the lowest elevation): what is the relief on the west (left) side of the river?   on the east (right) side of the river?

3. What is the highest point on the map, and what is its elevation?

4. Imagine that you are planning a *low-exertion* walk from the church across the river to the high point on the west side of this landscape. First on **Figure 23.2a** and then on **Figure 23.2b**, use a red colored pencil to draw the route you would take; give your reasons for choosing this easier route in terms of elevation change per distance traveled.

(a)

(b)

▲ **Figure 23.2** (a) Perspective view of a hypothetical landscape; (b) topographic map of that landscape. (After the U.S. Geological Survey)

5. On each figure use a blue pencil to draw in streams, using arrows to show the direction of stream flow. Note that there is a main stream in the valley and there are at least two tributary streams or creeks. How do the contour lines indicate the direction of flow?

SECTION 2

Constructing Contour Lines

Figure 23.3 presents elevation values for a selected landscape. The contour interval is 40 feet, and **index contours** (thicker lines—on this map every fifth contour, marked with the line's elevation) at 200 feet, 400 feet, and 600 feet are drawn for you in bold isolines. A river channel is noted on the map flowing along a valley separated from the ocean by a ridge. The elevation of the channel is marked at three locations.

Use Figure 23.3 to do the following work and answer the questions.

1. Using a pencil, sketch 40-foot-interval contour lines using the specific site elevations given. Note that we have drawn the 40-ft contour line for you. You must *interpolate* (estimate) elevations between known values. Draw your contour lines lightly at first as you determine the best portrayal, then darken in your work and erase stray pencil marks. (Do not worry if your lines go through the elevation labels.) Remember the contour line basics given in Section 1.

    The coastline is the datum—mean sea level. *Begin your work at the coast and work inland.* (*Hint:* Make this task easier by color-coding the elevations in 40-foot intervals. Circle all elevations from 0–40 feet in green, 41–80 feet in yellow, etc. Then, when drawing the contour lines, you will be "grouping" the colors, with the contour lines as separators between the color groups.)

2. Three vertical control bench marks, indicated by the letters "BM" and marked with an "x," are noted on the map. What are their elevations?

3. Three other specific spot elevations are noted with an "x." What are their elevations?

4. What effect do the intermittent (periodically dry) stream channels have on the topography of the region?

5. Given the trend and location of the 40-foot elevation contour, if you were walking along the water's edge, how would you characterize the topography and relief of the coastline? Steep? Gentle relief?

6. In what compass direction does the river flow? How can you tell? (See Section 1, question 5.)

7. Which portion (NW, NE, SE, or SW) of the map has the greatest relief?

▲ **Figure 23.3** A portion of the Stewarts Point quadrangle, enlarged and adapted with 200-, 400-, and 600-foot index contours highlighted.

8. **Challenge question**: If you were assigned the task of building a single general aviation runway 3700 feet in length, where would you place it? Draw your runway plan on the map to the correct scale. (Make it 100 feet wide.) Give your reasons for this site selection.

9. Google Earth™ activity, Stewarts Point quadrangle. For the KMZ file and questions go to mygeoscience-place.com. Then click on the cover of *Applied Physical Geography: Geosystems in the Laboratory.* Or visit the Mastering Geography Study Area.

SECTION 3

Topographic Profile

Relief refers to vertical elevation differences in a local landscape. In **Figure 23.4** of this exercise, you constructed contour lines and completed an analysis of a coastal landscape. The local topography included a coastal marine platform (terrace), low ridge, stream valley, and hill. The maximum relief on the map was approximately 700 feet, averaging about 400 feet.

An important method of topographic map analysis is construction of a **topographic profile**, the graphic representation of graduated elevations along a line segment drawn on a map. A **map view**, or plan view, is the normal way we view a map as if we were looking straight down from above ("bird's-eye" view). A cross section of a landscape—as if you sliced through the strata and obtained a side view—is a **profile view** and is demonstrated in Figure 23.4. The profile view shows you the "shape" of the land and demonstrates a line-of-sight perspective. As you work with topographic profiles, allow yourself to develop methods with which you are comfortable.

One method of preparing a profile is to draw a line connecting two points (a *transect*) along which you want to obtain the profile (Figure 23.4, Step 1). Take a piece of paper, fold the paper over and crease it for a stiff straight edge. The paper can be plain or graph paper. (Graph paper divided 10 squares to the inch is included at the back of this manual.) Lay the folded edge along the drawn line. Make tick marks at each point where a contour line makes contact with the edge of the paper and note the elevation at

that point. These will represent elevation points on the profile graph (Step 2). Carry the tick marks down to the graph at the appropriate elevation, plotting each point; connect the points with a smooth curved line to complete the profile (Step 3).

Construct your first topographic profile using **Figure 23.5**. Portrayed is an enlarged portion of the Palmyra, New York, quadrangle. The western New York region was blanketed during several advances of continental glaciers. The geomorphic feature you are profiling was deposited by a glacier.

A prepared graph is supplied with the figure; however, let's briefly discuss how it is set up. Note in Figure 23.5 that the horizontal scale of the map is 1:12,000, or 1.1″ = 1000′. If we used this same scale for the vertical axis on the topographic profile graph, the local relief of 192 feet would cover only 0.192 inch on the graph. To construct a readable and useful profile a technique of **vertical exaggeration** is employed. In this figure a vertical scale of 1:1200, or 1′ = 100′, is used and represents a 10 × (times) exaggeration of the horizontal scale. Other landscapes require different vertical exaggerations; the lower the maximum relief of a landscape, the greater the exaggeration should be to conveniently fit on a graph for analysis. For instance, the greater relief in **Figure 23.6** made an exaggeration of 3.3 × (times) more appropriate as compared to the 10 × exaggeration in Figure 23.5. The horizontal scale should be left at the same scale as the map.

Follow the procedure illustrated in Figure 23.5 to construct a profile along the line drawn in Figure 23.5, making tick marks at each point where the drawn line and contour lines intersect. Carry the elevations denoted by the tick marks down to the graph, plotting each point. The steeper the slope, the closer the points are placed; a gentle slope places them farther apart. The last step is to connect the points with a smooth curved line to visualize the relief and topography of this landscape. You may want to shade the area below the line to better display the profile.

▲ **Figure 23.4** Constructing a topographic profile using the edge of a piece of paper. (From Busch, Richard M., editor, *Laboratory Manual in Physical Geology*, 3rd ed., Macmillan Publishing Company © 1993.)

Use the profile constructed in Figure 23.5 to answer the following questions and completion items.

1. What is the horizontal distance of this topographic profile?

2. What is the maximum relief along this profile?

3. The landform feature you have profiled is a drumlin. Drumlins are formed by glacial deposits and are streamlined in the direction of the glacier's movement. They have a blunt end upstream, a tapered end downstream, and a rounded summit. Using your protractor, what direction was the glacier flowing?

4. Measure from the 500′ contour line on the west side of the feature, across the summit at 667′, to the 500′ contour line on the east side of the feature. How wide is the feature?

Derived from:

PALMYRA QUADRANGLE
NEW YORK—WAYNE Co.
7.5 MINUTE SERIES (TOPOGRAPHIC)

Huckleberry Swamp

N

550

600

550

500

500

550

650

667 X

600

500

500

Contour interval 10 feet
(Index contour every 50 feet)

Elevation (feet)
Vertical scale 1" = 100'
(vertical exaggeration 10x)

650

600

550

500

500

475'

470

0   1000   2000   3000   4000   5000   6000

1 mile

Surface distance (feet)
Horizontal scale : 1.1" = 1000'

▲ **Figure 23.5** Topographic profile from a portion of the Palmyra, New York, 7.5 minute series topographic quadrangle (enlarged).

▲ **Figure 23.6** Topographic profile from a portion of the Stewarts Point, California, 7.5 minute series topographic quadrangle (enlarged). From west to east along the profile, this landscape features a coastal marine terrace, a ridge, a river valley, and a hill rising to 731 feet above sea level. The coastline forms a contour line of equal elevation.

5. The horizontal scale is 1.1″ = _____′, so each horizontal square represents ___′. The vertical scale is 1″ = 100′, so each vertical square represents ___′. The total relief along the transect is ___′ How many vertical squares does it take to represent the total relief along the transect? If each vertical square represented as many feet as each horizontal square, so that the vertical scale was the same as the horizontal scale, how many vertical squares would it take to represent the maxium relief along the profile?_____

6. Google Earth™ activity, Palmyra quadrangle drumlin. For the KMZ file and questions go to mygeoscienceplace.com. Then click on the cover of *Applied Physical Geography: Geosystems in the Laboratory.* Or visit the Mastering Geography Study Area.

A second example of a vertical profile is in Figure 23.6. Using the same procedure as you did on the previous assignment, construct a topographic profile. Label the South Fork of the Gualala River (see Figure 23.4, Step 3), the bench mark at 112 feet, and various road crossings on your profile. The coastline forms a contour line of equal elevation—sea level. Note the change in contour interval and scale from that of Figure 23.5.
Use Figure 23.6 to answer further questions about preparing a topographic profile.

7. What is the length of this topographic profile in miles? in kilometers?

8. After you complete the topographic profile you'll notice a marine terrace, a relatively flat platform or "shelf," along the coast. You can find where the marine terrace ends by looking along the topographic profile for where the slope changes from flat to steeper. What contour elevation marks this change from the marine terrace to the hills inland? How wide is this terrace from the coast line to that contour line elevation in feet?

9. If we used a vertical scale identical to the horizontal scale, how many squares on the graph would accommodate the maximum relief along the profile? (Hint: use the same method you used in question 5 to find the answer.)

10. What is the relief (elevation difference between the highest and lowest points, in a local landscape) in the river valley along southwest slopes?

Along the northeast slopes?

11. Use your ruler to draw a line on the elevation profile in Figure 23.6. from the 731′ dot to the crest of the hill on the transect. This line represents your line of sight from the 731′ elevation. Lightly shade the locations that you cannot see from the 731′ elevation.

12. Can you physically see the shoreline when standing at the summit at 731 feet elevation? Explain.

Can you see the marine terrace from the summit?

Can you see the northwest end of the runway at Sea Ranch Airport (a point just south of the profile) from the summit? Explain.

SECTION 4

Interpreting Topographic Map Information

Topographic maps in the series of quadrangle maps prepared by the USGS, or Centre for Topographic Information in Canada, provide a wealth of information. Have your lab instructor show you an index map for your state or province that portrays all the quadrangle map coverage available. Your lab instructor will provide you with a topographic map sheet to examine as you follow along with this description of its elements.

The name of the topographic map quadrangle is given in the upper-right corner, with the state name and 15-minute or 7.5-minute series designation. If it is a 7.5-minute quad, it also signifies which portion of the published 15-minute quad it represents ("SW/4, NE/4," etc.). Other grids, in addition to latitude/longitude, are in use, so geographic coordinates of the **Universal Transverse Mercator (UTM)** grid system are listed near the corner margins of the quad map. The Public Lands Survey township and range system used in maps of the western United States is also noted.

COUNCIL BLUFFS NORTH QUADRANGLE
IOWA–NEBRASKA
7.5 MINUTE SERIES (TOPOGRAPHIC)

The lower-right corner carries the name of the quad; portion of 15-minute quad, if applicable; **geographic index number**; and the year of the map. The geographic index number is derived from the latitude and longitude of the lower-right corner of the map. Each corner also features complete geographic coordinates.

COUNCIL BLUFFS NORTH, IOWA—NEBR.
N4115–W9545/7.5

1956

PHOTOREVISED 1969 AND 1975

AMS 6866 IV SE—SERIES V876

The upper-left corner always carries the same credit line for the primary government agent for U.S. mapping.

UNITED STATES
DEPARTMENT OF THE INTERIOR
GEOLOGICAL SURVEY

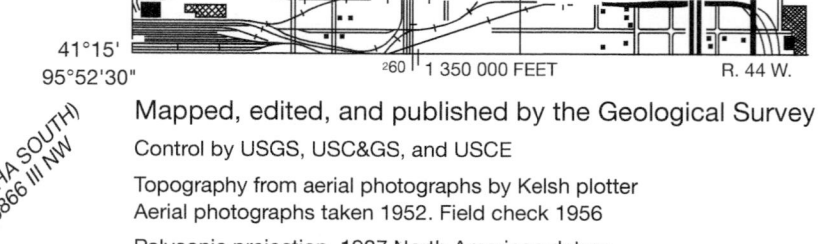

A more complete credit line and map preparation history is in the lower-left margin. If cooperative assistance from another entity was only minor, this may be listed in this place. A variety of such information might be featured in this label.

Mapped, edited, and published by the Geological Survey

Control by USGS, USC&GS, and USCE

Topography from aerial photographs by Kelsh plotter
Aerial photographs taken 1952. Field check 1956

Polyconic projection. 1927 North American datum
10,000-foot grid based on Iowa coordinate system, south zone
1000-metre Universal Transverse Mercator grid ticks,
zone 15, shown in blue

Red tint indicates area in which only landmark buildings are shown

Revisions shown in purple compiled from aerial photographs
taken 1969 and 1975. This information not field checked

Purple tint indicates extension of urban areas

Where applicable, the names of adjoining topographic quadrangles are given in parentheses at each corner and along the sides. If this reference appears, you can assume that the map name refers to a quad at the same scale and the same series.

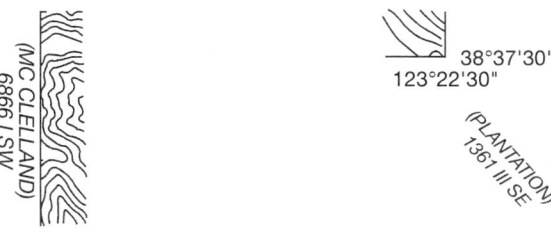

Earth's geographic North Pole and magnetic North Pole locations do not coincide. In addition, the magnetic pole slowly migrates from year to year. The lower left-center margin features a magnetic declination diagram for the year of the map field survey or date of map revision. This declination is taken from an **isogonic map** (Figure 3.3) that maps magnetic declination. This diagram allows you to correct compass bearing readings when using the topographic map. A state map graphically shows the location of the quad in the lower right-center margin. Here are some samples of this designation:

The map scale (representative fraction and graphic or bar scale), contour intervals, any supplementary contour intervals, vertical datum ("National Geodetic Vertical Datum of 1929" appears on post-1975 maps), depth or sounding information in bodies of water, and sales information are shown in the lower-center margin of the map sheet.

SCALE 1:24 000

CONTOUR INTERVAL 10 FEET
NATIONAL GEODETIC VERTICAL DATUM OF 1929

A complete map symbol key should be used for interpreting the map (inside front cover of this lab manual), although road symbols used for the quad are shown in the lower-right corner.

1200

INTERIOR—GEOLOGICAL SURVEY, RESTON, VIRGINIA—1977

4570000m.N.

41°15'

269000m.E.

95°45'

(MINEOLA)
6866 II NW

ROAD CLASSIFICATION

Heavy-duty___ _____   Light-duty___ _____

Unimproved dirt = = = = = = =

U.S. Route   State Route

Interstate Route

Use a topographic map *provided by your lab instructor* to answer the following questions and completion items.

1. Name, state, and series of the map:

2. If a 7.5-minute map, give its location and position relative to the 15-minute series quad for this region:

3. Did any other agencies or entities participate in the preparation of this map?

4. List the geographic index number for this quad:

5. Denote the latitude and longitude of each corner of the quad map:

   NE corner:

   SE corner:

   SW corner:

   NW corner:

6. List the nearest UTM (Universal Transverse Mercator) coordinate for the upper-right corner of the map:

7. Describe the complete map history and credit line:

8. If you need to obtain the maps adjoining this quadrangle, what are their names? Label these names along the appropriate sides and corners on the quadrangle.

9. What is the representative fraction scale for this map?

   Convert this to a written (verbal) scale. (Show your work.)

10. Using the graphic or bar scale on the map, determine the distance between two points selected
    by your instructor. The points are: and
    Distance in kilometers   ; and miles

11. Calculate the total area of the map in square kilometers. (Show your work.)

    Now calculate the area in square miles. (Show your work.)

12. What is the magnetic declination for your quadrangle map?

    Which direction (and how many degrees) would you have to turn in order to adjust for this declination?

    To what azimuth would your compass needle point?

    Hikers and backpackers frequently include topographic maps and compasses among their equipment.
    Magnetic declination must be taken into account when using a compass, and adjustments must be
    made. To do this, simply superimpose the magnetic compass points (based on the declination arrow)
    over the true north compass points (N, E, W, S). It is easy to see that, if the magnetic declination is 10°E,
    you must turn 10° west, so that your compass north arrow is pointing to 350° (360°−10° = 350°).

13. List any additional information provided in the margins of this particular map quadrangle:

    Your instructor may want to provide you with additional questions pertaining to the topographic map
    that has been selected for this activity. Questions may include features specific to the chosen map and
    use of symbols for which you might need to consult the topographic map legend on the inside front
    cover of this manual.

Mastering Geography

Looking for additional lab prep and review materials? Visit the Study Area in Mastering Geography to enhance your
geographic literacy and spatial reasoning skills by accessing MapMaster™ interactive maps, videos, *Mobile Field Trips,
Project Condor* Quadcopter Videos, *In the News* articles, flashcards, web links, self-study quizzes, and other resources.
Contact your professor to find out how you can access Mastering Geography.

NOTES

Name: _____   Laboratory Section: _____

Date: _____   Score/Grade: _____

Video
Exercise 24
Pre-Lab Video

http://goo.gl/JZKYs3

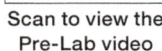

Scan to view the
Pre-Lab video

LAB EXERCISE

24

Topographic Analysis: Fluvial Geomorphology

One of the benefits you receive from studying physical geography is an enhanced appreciation of the scenery. As we visit Earth's varied landscapes, we witness the active physical processes that sculpt the land. *Endogenic processes* produce internal forces within our planet that build landforms. However, as the landscape is formed, a variety of *exogenic processes* produce external forces that simultaneously wear these features down. The landscape we see is the product of this continuous struggle of building and destruction.

Geomorphology is a science that analyzes and describes landforms—their origin, evolution, form, and spatial distribution. Geomorphology is an important discipline within physical geography.

Denudation is a general term that refers to all processes that cause reduction or rearrangement of landforms. The principal denudation processes affecting surface materials include *weathering, mass movement, erosion, transportation*, and *deposition*, as produced by the agents of moving water, air, waves, and ice, and the pull of gravity.

Fluvial (moving water; overland flow and streams) erosion processes and their resulting landforms dominate land surfaces throughout the world. Specific geomorphic landforms and processes resulting from fluvial actions are analyzed in this exercise using topographic maps and illustrations. Lab Exercise 24 features mobile-friendly QR links to Google Earth virtual tours that fly you to the topographic map location, as well as optional online Google Earth activities with downloadable media and worksheets at mygeoscienceplace.com and the Mastering Geography Study Area.

Key Terms and Concepts

alluvium
anticline
backswamp
denudation
drainage basins
drainage patterns

floodplain
fluvial
geomorphology
natural levees
oxbow lakes

syncline
watershed
water gap
wind gap
yazoo tributary

KEY**LEARNING**concepts

After completion of this lab, you should be able to:

- *Analyze* and *describe* fluvial geomorphic processes using topographic maps, photos, and Google Earth™ imagery.
- *Interpret* several fluvial processes and the characteristic landscapes that are produced, including drainage basins and floodplains.
- *Analyze* fluvial processes in a folded landscape, Cumberland Gap, Maryland.

Materials/Sources Needed

pencil
ruler

calculator
Google Earth

Lab Exercise and Activities

SECTION 1

Drainage Patterns and Floodplains

Stream-related processes are termed **fluvial** (from the Latin *fluvius*, meaning "river"). Fluvial systems exhibit characteristic processes and produce predictable landforms. The erosive action of flowing water and the deposition of stream-transported materials produce landforms. A stream system behaves with randomness, unpredictability, and disorder as described by the phenomena of chaos. Yet, enough regularity exists to allow some predictability.

A stream is a mixture of water and solids—carried in solution, by suspension, and by mechanical transport. **Alluvium** is the general term for the clay, silt, and sand transported and then deposited by running water.

Streams may drain large regions. Consider the travels of rainfall in north-central Pennsylvania. This water feeds hundreds of small streams that flow into the Allegheny River. At the same time, rainfall in southern Pennsylvania feeds hundreds of streams that flow into the Monongahela River. The two rivers then join at Pittsburgh to form the Ohio River. The Ohio flows southwestward and at Cairo, Illinois, connects with the Mississippi River, which eventually flows on past New Orleans into the Gulf of Mexico. Each contributing tributary, large or small, adds its discharge and sediment load to the larger river.

Streams are organized into areas or regions called **drainage basins**. A drainage basin is the spatial geomorphic unit occupied by a river system, defined by ridges that form *drainage divides*, i.e., the ridges are the dividing lines that control into which basin precipitation drains (**Figure 24.1**).

▲ **Figure 24.1** The drainage basin and its watershed, separated from other basins by a drainage divide.

Drainage divides define a **watershed**, the catchment area of the drainage basin. A major drainage basin system, such as the one created by the Mississippi-Missouri-Ohio river system, is made up of many smaller drainage basins, which in turn comprise even smaller basins, each divided by specific watersheds. Surface runoff concentrates in *rills*, or small-scale indentations, which may develop further into *gullies* and *stream courses* with distinct drainage patterns. The resultant **drainage pattern** is an arrangement of channels determined by slope, differing rock resistance, climatic and hydrologic variability, and structural controls imposed by the landscape.

Figure 24.2 shows seven common drainage patterns. A most familiar pattern is *dendritic drainage* (a). This treelike pattern (from the Greek word dendron, meaning "tree") is similar to that of many natural systems, such as capillaries in the human circulatory system, or the vein patterns in leaves, or tree roots. Energy expended by this drainage system is efficient because the overall length of the branches is minimized. Dendritc patterns form in regions with v-shaped stream-out valleys of Impermeable rock. The *trellis* drainage pattern (b) is characteristic of dipping or folded topography, which exists in the nearly parallel Ridge and Valley Province of the eastern United States, where drainage patterns are influenced by rock structures of variable resistance and folded strata. We see this later in this lab, in the Cumberland topographic map in Section 3.

The sketch in Figure 24.2b suggests that the headward-eroding part of one stream could break through a drainage divide and *capture* the headwaters of another stream in the next valley, and indeed this does happen. The sharp bends in two of the streams in the illustration are called *elbows of capture* and are evidence that the stream has breached a drainage divide. This type of capture, or *stream piracy*, occurs in other drainage patterns.

The remaining drainage patterns in Figure 24.2 are caused by other specific structural conditions. A radial drainage pattern (c) results from streams flowing off a central peak or dome, such as occurs on a volcanic mountain. We will see this later in the Mt. Rainier topographic map in Lab Exercise 25, Section 2. *Parallel* drainage (d) is associated with steep slopes. A *rectangular* pattern is formed by a faulted and jointed landscape, directing stream courses in patterns of right-angle turns (e). Structural domes, with concentric patterns of rock strata guiding stream courses, produce *annular* patterns (f). In areas having disrupted surface

▲ **Figure 24.2** Seven most common drainage patterns [Based on A. D. Howard, "Drainage analysis in geological interpretation: A summation," *Bulletin of American Association of Petroleum Geologists* 51 (1967), p. 2248.]

patterns, such as the glaciated shield regions of Canada and northern Europe, a *deranged* pattern (g) is in evidence, with no clear geometry. This will be seen in the Jackson, Michigan, topographic map in Lab Exercise 25, Section 3.

The low-lying area near a stream channel that is subjected to recurrent flooding is a **floodplain**, formed when the river leaves its channel during times of high flow. Thus, when the river channel changes course or when floods occur, the floodplain is inundated with water. When the water recedes, alluvial deposits generally mask the underlying rock. Episodic flood events are often devastating, as we witnessed during the Midwest floods of summer 1993; Ohio River floods of late winter 1996–1997; and Red River floods that drowned Fargo and Grand Forks, North Dakota, and the regions surrounding Winnipeg, Manitoba, in spring 1997; or North Carolina in 1999, following the landfall of three hurricanes. **Figure 24.3** illustrates a characteristic floodplain, with the present river channel embedded in the plain's alluvial deposits.

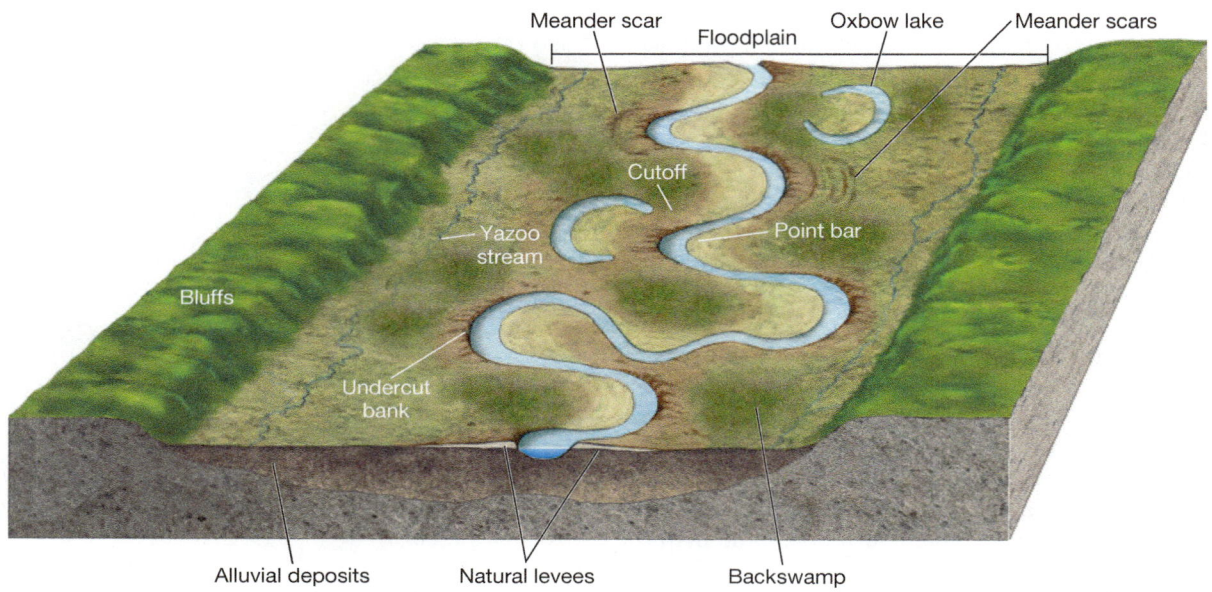

▲ **Figure 24.3** Typical floodplain landscape and related geomorphic features

On either bank of most streams, **natural levees** develop as by-products of flooding. When flood waters arrive, the river overflows its banks, loses velocity as it spreads out, and drops a portion of its sediment load to form the levees. Notice on Figure 24.3 an area labeled **backswamp** and a stream called a **yazoo tributary.** The natural levees and elevated channel of the river prevent this yazoo tributary from joining the main channel, so it flows parallel to the river and through the backswamp area. When a meandering stream erodes its outside bank as the curve migrates downstream, the neck of land created by the looping meander eventually erodes through and forms a *cutoff.* When the former meander becomes isolated from the rest of the river, the resulting **oxbow lake** may gradually fill in with silt or may again become part of the river when it floods.

Animation
Stream
Processes, Flood
Plains

https://goo.gl/mlH2m7

Drainage Patterns and Drainage Basins

Figures 24.4 through **24.7** show some of the different types of drainage basins. Use Figure 24.2 to help determine which type of pattern is shown in each figure.

▲ **Figure 24.4**

1. Which type of drainage pattern is shown in Figure 24.4? Briefly describe the structural conditions that have created this pattern.

121.90°W   121.72°W

40.02°N   40.02°N

39.88°N   39.88°N

121.90°W   121.72°W

2   1   0   2
MILES

2   1   0   2
KILOMETERS

▲ Figure 24.5

2. Which type of drainage pattern is shown in **Figure 24.5**? Briefly describe the structural conditions that have created this pattern.

122.26°W   122.13°W

41.46°N — 41.46°N

41.36°N — 41.36°N

122.26°W   122.13°W

2   1   0   2

MILES

2   1   0   2

KILOMETERS

▲ Figure 24.6

3. Which type of drainage pattern is shown in **Figure 24.6**? Briefly describe the structural conditions that have created this pattern.

119.55°"W   119.26°W

37.85°N   37.85°N

37.62°N   37.62°N

119.55°W   119.26°W

```
4        2        0              4
```
MILES
```
4      2        0          4
```
KILOMETERS

▲ **Figure 24.7**

4. Which type of drainage pattern is shown in **Figure 24.7**? Briefly describe the structural conditions that have created this pattern.

Philipp, Mississippi, Quadrangle

Figure 24.8 is a portion of the Philipp, Mississippi, 7.5-minute quadrangle in northwestern Mississippi. The Tallahatchie River flows through the landscape from west to east. (Remember to use the topographic map symbol legend inside the front cover of this manual.)

Review Figure 24.4, analyze Figure 24.8 closely, and answer the following questions and completion items about Figure 24.8.

Virtual Tour
Philipp, MS

https://goo.gl/q7ixc4

5. What is the contour interval on this map segment?

Given this value, determine the maximum relief shown:

What is the highest spot elevation ("X") on the map and in what PLSS (Public Land Survey Section) section is it located?

6. What is the name of the primary meridian that was used to survey this region for the Public Land Survey System?

7. What is the elevation of the water surface of the lake north of the Jones Chapel Cemetery?

8. What type of geographic feature is the water body called Fish Lake, and how did it form? Use the illustration in Figure 24.3 to identify this water feature.

9. With a dark blue or black colored pencil or pen, outline all the water features that were formed by the same process as Fish Lake. What is the name of the widest of these water features on the map?

10. There is a stream connecting Possum Bayou with the smaller lake called Fish Lake. With a dark blue or black colored pencil or pen, draw an arrow indicating which direction the water is flowing.

11. Do the fluvial features have any effect on road patterns? Explain and give some examples. Highlight your examples with a colored pencil or pen on the topographic map.

PHILIPP, MS

MISSISSIPPI

QUADRANGLE LOCATION

Matthews Bayou

Saint Paul Cem

Hampton Lake

Possum Bayou

Fish Lake

Point

Fish Lake

Tallahatchie River

Pecan

River Rd

T23N R1W

TALLAHATCHIE CO
LEFLORE CO

T22N R1W

Shegogue Lake

Bellchase Rd

Jones Chapel Cem

Bellchase Rd

T23N R1E

Pecan
Point Cut-Off

T22N R1E

▲ Figure 24.8

MILES

FEET

KILOMETERS

CONTOUR INTERVAL 5 FEET

SCALE 1: 27,000

12. Would you go fishing in the area labeled Shegogue Lake in LeFlore County? What kind of feature is the area labeled Shegogue Lake in LeFlore County? Circle or highlight at least two other areas on the map that are the same type of feature.

13. In the space below, sketch the topographic map symbol for a marsh or swamp. With a blue colored pencil or pen, lightly shade the backswamp areas on the topographic map.

14. Optional Google Earth™ activity, Drainage patterns. For the KMZ file and questions go to mygeoscienceplace.com. Then click on the cover of *Applied Physical Geography: Geosystems in the Laboratory.* Or visit the Mastering Geography Study Area.

SECTION 2

Sacramento River, California

al Tour
·mento
ver

o.gl/r01HNo

Streams often are used as natural political boundaries, but it is easy to see how disagreements might arise when boundaries are based on river channels that shift around. For example, the Ohio, Missouri, and Mississippi Rivers can shift their positions quite rapidly during times of flood, thus creating a mismatch between the boundary and river.

The boundary separating Glenn and Butte Counties in California provides us with an example, as shown in **Figure 24.9**. The Glenn–Butte border was originally placed midchannel in the meandering Sacramento River. The river has shifted course, but the state boundary line still follows the former channel meander, placing parts of Glenn County on the same side of the river as Butte County and vice versa!

Figure 24.9 is a portion of the Ord Bend 7.5-minute quadrangle, in northern California. The Sacramento River flows through the landscape from north to south.

Analyze Figure 24.9 and answer the following completion items.

1. What is the contour interval on this map segment?

   Given this value, determine the relief from Dunning Slough in the north to Perch Slough in the south.

2. Use a piece of string and the scale bar on the map to measure the river from the north end of the map to the south end. How many miles did you measure? What is the straight-line distance from the river at the north end of the map to the south end in miles?

The ratio of river distance to straight-line distance is the sinuousity of the river. For straight streams it is less than 1.05, for winding and twisty streams is is between 1.05 and 1.5, for meandering streams, it is greater than 1.5. What value would you expect to find, based on the type of river this is? What value did you find?

3. Shade in areas on Figure 24.9 north of 39° 43′ on the Glenn County (west) side of the river but legally in Butte County and vice versa.

4. Find Murphy Slough and Jenny Lind Bend. Which type of feature are they? Use a green colored pencil or pen and color in the other features of that type on the map.

5. Use a red colored pencil or pen and outline at least two meander scars on the map.

6. Which set of features was formed first, and which set is older: those named in question 4 or question 5? Briefly explain how those features are formed by the river. Begin with the river.

7. Why do you think Phelan Island and Golden State Island have the names they do? Describe how they were formed.

8. From Golden State Island downstream to Shannon Slough, draw arrows on the cut bank side of the river to indicate the direction that the river will migrate initially. Use a dark blue colored pencil or pen and shade in the point bar region where deposition is occurring now.

9. In the space below, make a sketch showing how the meander bend will become a cutoff, an oxbow lake, and a meander scar.

▲ Figure 24.9

▲ Figure 24.10

SACRAMENTO RIVER, CA

CALIFORNIA

QUADRANGLE LOCATION

River channels

1896
1908
1923
1935
1937
1946
1955
1956
1960
1964
1969
1976
1981
1991
1997
2012

0   0.5   1   2 Miles

0   2,500   5,000   10,000 Feet

0   0.5   1   2 Kilometers

CONTOUR INTERVAL 10 FEET   SCALE 1: 48,000

Figure 24.11 is a rendering of a digital elevation model (DEM) of the same region as the topographic map of the Sacramento River in Figure 24.9. DEMs sometimes reveal hidden features. Figures 24.9, 24.10, and 24.11 are printed so that you can view both 24.9 and 24.11 at the same time.

10. Locate and identify several features on Figure 24.10 that are not obvious on the topographic map in Figure 24.9. Use a dark pen or pencil to label them clearly. Locate and label several features on Figure 24.11 that you can see on both the topographic map and the DEM.

Figure 24.10 is a rendering of historic channels of the Sacramento River from 1896 to 2012 placed on aerial imagery. Review Figures 24.9, 24.10, and 24.11 and answer the following questions.

11. The county boundary between Butte and Glenn counties around Jenny Lind Bend clearly does not follow the current river channel. When was the last year that the channel and the county boundary were in alignment?

12. Describe the general pattern of migration of the meander bends. Do they migrate upstream? downstream? Provide a specific example or examples.

13. Optional Google Earth™ activity, Sacramento River and political boundaries. For the KMZ file and questions go to mygeoscienceplace.com. Then click on the cover of *Applied Physical Geography: Geosystems in the Laboratory*. Or visit the Mastering Geography Study Area.

SECTION 3

Cumberland Quadrangle

In folded landscapes, ridges and valleys are the prominent *topographic features*, reflecting the **anticlines** and **synclines** that are the *structural landscape controls*. Anticlines are the upfolds, and synclines are the downfolds. In the Ridge and Valley geomorphic province of the Appalachians, **water gaps** occur where a river cuts through an *anticlinal* ridge (**Figure 24.12**). Such a geomorphic feature is thought to form when a preexisting stream course flows across a landscape with buried geologic structures. As the overburden cover is removed, the downcutting stream cuts into the underlying structure. The stream valley is described as *superposed* where Wills Creek cuts through an anticlinal ridge (Haystack Mountain and Wills Mountain) in this example from Cumberland, Maryland. The importance of these water gaps to early transportation and migration is

obvious. Note the abandoned canal in the southeast portion of the map.

Wind gaps are passes that were formerly occupied by a stream that subsequently was captured and incorporated into another stream drainage. These abandoned stream valleys are generally dry. Two remarkable examples of these features are on the portion of the Cumberland 7.5-minute quadrangle presented as **Figure 24.13**.

Refer to Section 1 of this lab exercise and the review of drainage basins and drainage patterns. Viewing Figure 24.12, it is easy to imagine the direction in which water would flow down the anticlines into streams that occupy the synclines, establishing the trellis drainage pattern in these ridge and valley regions. Keep this image and the trellis drainage pattern in mind. Analyze Figure 24.13 to answer the following questions and completion items.

Virtual Tour
Cumberland, MD

https://goo.gl/k3XtYQ

▲ Figure 24.11

SACRAMENTO
RIVER, CA

CALIFORNIA

QUADRANGLE LOCATION

0	0.5	1	2 Miles
0	2,500	5,000	10,000 Feet
0	0.5	1	2 Kilometers

CONTOUR INTERVAL 10 FEET   SCALE 1: 48,000

▲ **Figure 24.12** The Susquehanna River in Pennsylvania, a superposed stream. The Susquehanna River established its course on relatively uniform rock strata that covered more complex geologic structure below. Over time, as the landscape eroded, the river "superposed" its course onto the older structure by cutting through the resistant strata. [*Landsat-7*, NASA.]

1. What is the contour interval on this map segment?   What is the scale of this map?

2. Wills Creek at The Narrows is a *superposed stream* forming a *water gap*. How wide is the water gap gorge (distance between the 1560-foot contour to the north and the 1480-foot contour to the south)?

   How deep is the water gap at The Narrows?

3. What is the relief between the floor of the wind gap that Braddock Road runs through and the floor of the valley to the north in which the stream Braddock Run flows?

▲ Figure 24.13

A wind gap was formed when the lower course of Braddock Run was diverted (captured) by a tributary of Wills Creek (stream in the water gap). How wide is the wind gap, measured from the 1400′ contour line on the north rim to the 1500′ contour line on the south rim?

How deep is the wind gap, measured from the Forest Drive on the north rim to Braddock Road in the gap?

4. What is the elevation of the floor of the wind gap? What is the elevation of the valley to the north of the wind gap (the same valley that Braddock Run and Wills Creek run through)?

What is the elevation of the valley to the south of the wind gap (the valley with the Allegheny County/ Mineral County lines)? What is the vertical relief between the wind gap and the valley to the north? What is the vertical relief between the floor of the wind gap and the valley to the south?

5. Imagine a topographic profile from west of Narrows Park, along Braddock Run at the 700-foot contour line, to the 760-foot contour line south of Allegany High School at the southeast corner of the cemetery. How long is this profile line in straight distance?

What is the highest point along the profile?

What is the relief from Allegany High School to the highest point along the profile (south of the water gap)?

6. What topographic features appear to be controlling the stream course of the north branch of the Potomac River (joins Wills Creek along the southeast portion of the map), and what are the names of the structural landscape controls?

7. Optional Google Earth™ activity, Cumberland Gap. For the KMZ file and questions go to mygeoscience-place.com. Then click on the cover of *Applied Physical Geography: Geosystems in the Laboratory*. Or visit the Mastering Geography Study Area.

Mastering Geography

Looking for additional lab prep and review materials? Visit the Study Area in Mastering Geography to enhance your geographic literacy and spatial reasoning skills by accessing MapMaster™ interactive maps, videos, *Mobile Field Trips, Project Condor* Quadcopter Videos, *In the News* articles, flashcards, web links, self-study quizzes, and other resources. Contact your professor to find out how you can access Mastering Geography.

NOTES

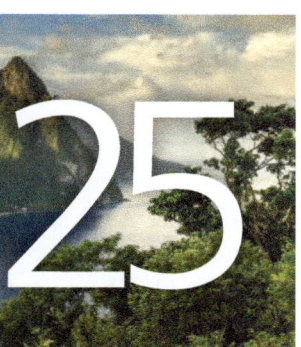

LAB EXERCISE

25 Topographic Analysis: Glacial Geomorphology

About 77% of Earth's freshwater is frozen, with the bulk of that ice in just two places—Greenland and Antarctica. The remaining ice covers various mountains and fills some alpine valleys. More than 32.7 million cubic kilometers (7.8 million cubic miles) of water is tied up as ice.

A **glacier** is a large mass of ice, resting on land or floating as an ice shelf in the sea adjacent to land. Glaciers are not frozen lakes or groundwater ice. Instead, they form through the continual accumulation of snow that recrystallizes under its own weight into an ice mass.

Glacial erosion and deposition produce distinctive landforms that differ greatly from the way the land looked before the ice came and went. Worldwide, glacial ice is in retreat, melting at rates exceeding anything in the recent ice record. In the European Alps

alone, some 75% of the glaciers have receded in the past 50 years, losing more than 50% of their ice mass since 1850. At this rate, the European Alps will have only 20% of their preindustrial glacial ice left by 2050.

In this lab exercise, topographic map interpretation continues. Specific exogenic geomorphic processes and the resulting landforms from glacial (moving ice) action are analyzed using topographic and contour maps, aerial imagery, and illustrations. A complete legend to topographic map symbols is inside the front cover of this lab manual. Lab Exercise 25 features mobile-friendly QR links to Google Earth virtual tours that fly you to the topographic map location, as well as optional online Google Earth activities with downloadable media and worksheets at mygeoscienceplace.com and the Mastering Geography Study Area.

Key Terms and Concepts

alpine glacier	esker	outwash plain
arêtes	glacier	paternoster lakes
cirque	horn	tarn
col	kame	terminal moraine
continental glacier	kettle	till plains
drumlin	medial moraine	valley glacier
end moraine	moraine	

KEY**LEARNING**concepts

After completion of this lab, you should be able to:

- *Analyze* and describe geomorphic processes by using topographic maps and Google Earth™ imagery.

- *Interpret* and *describe* several glacial processes and the characteristic landscapes that result, including erosional and depositional features of alpine and continental glaciers.

Materials/Sources Needed

pencil	string
ruler	calculator
Google Earth	

Lab Exercise and Activities

SECTION 1

Glaciated Landscapes

With few exceptions, a glacier in a mountain range is called an **alpine glacier**, or *mountain glacier*. The name comes from the Alps, where such glaciers abound. On a larger scale than individual alpine glaciers, a continuous mass of ice is known as a **continental glacier** and in its most extensive form is called an *ice sheet*. Two additional types of continuous ice cover associated with mountain locations are designated as *ice caps* and *ice fields*.

Alpine glaciers form in several subtypes. One prominent type is a **valley glacier**, an ice mass constricted within a valley that was originally formed by stream action. As a valley glacier flows slowly downhill, the mountains, canyons, and river valleys beneath its mass are profoundly altered by its passage. Most alpine glaciers originate in a mountain snowfield that is confined in a bowl-shaped recess. This scooped-out erosional land form at the head of a valley is called a **cirque**. In the cirques where the valley glaciers originated, small mountain lakes, called **tarns,** may form after the valley glacier fully melts. Some cirques may contain small, circular, stair-stepped lakes, called **paternoster** ("our father") **lakes**, for their resemblance to rosary (religious) beads. As the cirque walls erode away, sharp ridges form, dividing adjacent cirque basins. These **arêtes** ("knife edge" in French) become the sawtooth, serrated ridges in glaciated mountains. Two eroding cirques may reduce an arête to a saddle-like depression or pass, called a **col**. A **horn** (pyramidal peak) results when several cirque glaciers gouge an individual mountain summit from all sides. The deposition of glacial sediment produces a specific landform, a **moraine**. Several types of moraines are associated with alpine glaciation. A lateral moraine forms along each side of a glacier. If two glaciers with lateral moraines join, a **medial moraine** may form. Other moraines are associated with both alpine and continental glaciation. Eroded debris that is dropped at the glacier's farthest extent is a **terminal moraine**. **End moraines** may also be present, formed at other points where a glacier paused after reaching a new equilibrium between accumulation and ablation. **Figure 25.1** shows a cross section of a typical retreating alpine glacier and **Figure 25.2** shows a typical post-glaciated landscape.

Mount Darwin, California, Quadrangle

Virtual Tour
Mt. Darwin

https://goo.gl/nvFWWd

Analyze **Figure 25.3**. The NAIP photos were taken in 2014.

1. A cirque basin may serve as an accumulation zone for alpine glaciers. Using contour lines on the mountains and on the glaciers, how many ice-filled cirques (scooped out amphitheater-like basins) do you identify on the map?   What is the elevation of the highest glacial ice surface on the map?

2. What type of glacial feature is the unnamed peak, with an elevation of 13,253 ft, north of Mount Darwin?

▲ **Figure 25.1** Cross section of a typical retreating alpine glacier

3. What is the relief (elevation difference) in feet between Mount Goethe and Goethe Lake?

4. **a)** What is the linear distance in feet between Mount Goethe and the south shoreline of Goethe Lake?

   b) Take the relief in feet that you determined in Question 3 and divide it by the distance between the two locations in miles (divide the distance you found in 4 a) by 5280 to find the distance in miles). In terms of slope, state this distance in feet per mile: (Show your work.)

   c) Take the relief in feet that you determined in Question 3 and divide it by the distance between the two locations in feet from 4 a) to determine the percent grade (relief divided by distance). What percent grade did you determine? (Show your work.)

5. Locate Sky High Lake in Section 34 on the topo map. What type of feature is Sky High Lake? Use a blue pencil or pen to circle two similar features on Figure 25.3.

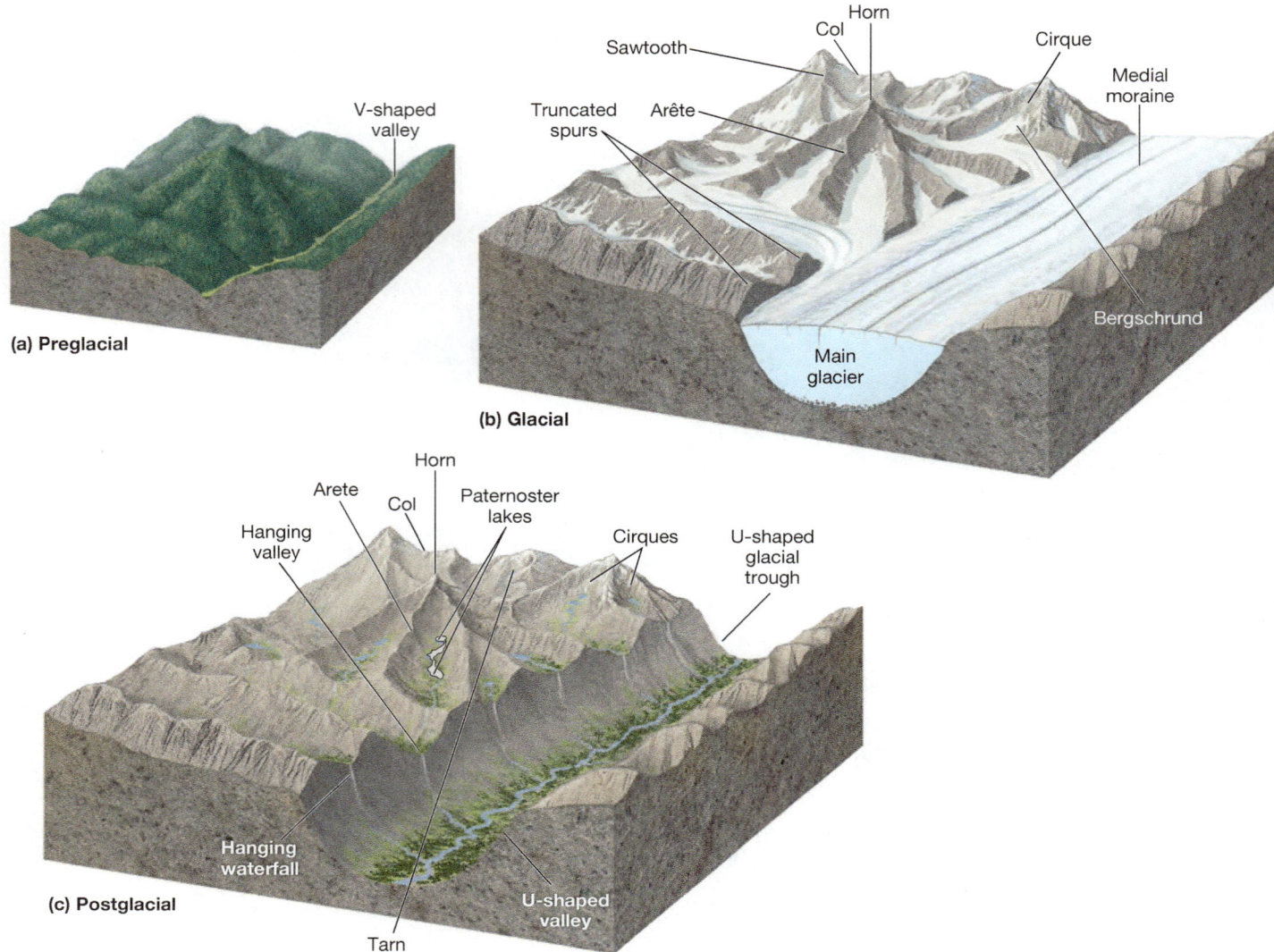

▲ **Figure 25.2** The work of alpine glaciers, during and after glaciation. (Based on W.M. Davis)

6. What is the term for the string of lakes in Darwin Canyon?

7. South of Goethe Lake find Mount Goethe on Figure 25.3 (topographic map) and **Figure 25.4** (NAIP photo).

   a) How many glaciers and moraine fields are there north of the red line that indicates the Wilderness Boundary that runs from the west edge of the map to Mt. Goethe and Muriel Peak? Count features along the length of the boundary across the bottom of Sections 31 (not labeled), 32, and 33. How many are there to the immediate south of that same Wilderness Boundary line?

▲ Figure 25.3

    b) Compare and contrast the number of glaciers and moraine fields that lie on either side of the National Forest and Wilderness Boundary (also labeled National Park Boundary) that runs roughly north and south from Mt. Darwin to the top of the map. East-facing slopes receive sunlight in the morning when the air is cooler, and west-facing slopes receive sunlight in the afternoon when the air is warmer. Would you expect to find more on the west-facing or east-facing slopes? How many did you find on each side?

    c) Describe the orientation of the slopes with snow on them on the topographic map and the NAIP photo. Is there more snow on north- facing slopes or on south-facing slopes? Which slope receives more direct sunlight? How would this affect the amount of glaciation that would occur on each slope? On which slope would you expect to find more glaciers in the Southern Hemisphere, north-facing or south-facing slopes? Briefly describe the relationship between slope orientation, snow accumulation, glacier formation, and glacial abrasion.

8. a) What term would you use to describe the ridge west of Mount Goethe?

    b) What is the term for the feature found on the ridge between Mount Goethe and Muriel Peak?

9. Optional Google Earth™ activity, Mount Darwin, California. For the KMZ file and questions go to mygeoscienceplace.com. Then click on the cover of *Applied Physical Geography: Geosystems in the Laboratory*. Or visit the Mastering Geography Study Area.

SECTION 2

Alpine Glacier and Mount Rainier Quadrangle

Virtual Tour
Mt. Rainer

https://goo.gl/6ezfCQ

An alpine (valley) glacier has profound effects on the landscape. Figures 25.1 and 25.2 illustrate views during and after a glacier's passage. A typical river valley has a characteristic **V**-shape and stream-cut tributary valleys. During active glaciation, glacial erosion and transport actively remove much of the regolith (weathered bedrock) and the soils that covered the stream valley landscape. When climates warm and the ice has retreated, the valleys are exposed as **U**-shaped glacially carved valleys, greatly changed from their previous stream-cut form. You can see the oversteepened sides and the straightened course of the valley.

Figure 25.5 shows the contour lines for an alpine glacier scene with key portions labeled. Please compare this to the alpine glaciation illustrations in Figures 25.1 and 25.2. Mount Rainier is featured in **Figure 25.6**—a dormant composite volcano in the Cascade Range that is covered by several glaciers and glacial features.

▲ **Figure 25.4** National Aerial Imagery Program image of the Mount Darwin area of California.

CONTOUR INTERVAL 200 FEET

▲ **Figure 25.5** Alpine glacier landscape
Contour map of an alpine glacial landscape.

Activities and completion items related to alpine glacier geomorphology:

1. Figure 25.4 shows a contour map of a hypothetical landscape that has been shaped by alpine glaciation. As glaciers carved their way down preexisting V-shaped stream valleys, they widened the valleys into U-shaped valleys separated by sharp ridges. Hanging valleys/troughs left stranded above the main valley floor will be occupied by streams that form picturesque waterfalls as they plunge over the steep edge.

   Glaciers may scour out the bedrock of the valley floor, leaving depressions that, as the glacier retreats and a stream reoccupies the valley, subsequently fill with water, forming glacial lakes called tarns. Steep-walled cirques mark the origin of the glaciers high on the mountain slope. Keep this contours map, along with the Mount Darwin topo map and aerial imagery, in mind when analyzing the topographic map of Mount Rainier.

   After reviewing Figures 25.1 and 25.2, answer the following questions about Figure 25.5:

   If you wanted to go swimming in a tarn, which letters would you look for?

   If you were a rock climber and wanted to scale the steep sides of the valley, which letters would you look for?

   Which letters indicate the location of a pass on the ridge between two valleys?

   If you wanted to climb the horn created by three glaciers, which letter would you look for?

   What is the elevation of the peak?

   Find and label the following features on Figure 25.5: a hanging valley; the U-shaped valley formed by the main glacier; a cirque; a horn; an arête (the narrow ridge that separates two valleys); a col (a pass on an arête); and a chain of paternoster lakes (if they were filled with water).

2. Refer to the Figure 25.6 Mount Rainier topographic quadrangle. What is the scale of this map as a representative fraction (1:???)? as a written scale (one inch equals ??? miles)?

   What is the contour interval of the topo map?

3. Name several glaciers on which you identify medial and lateral moraines. Use a brown colored pencil or dark pen and color down the middles of at least three medial moraines.

4. What type of feature is Sunset Amphitheater?

▲ Figure 25.6

5. What type of feature is labeled Nisqually Cleaver? List several other features that are the same type, including at least two that are not named Cleaver.

6. The summit of Mount Rainier is at 14,410 ft (4392 m). What is the lowest elevation shown on this map?

   Where is this location? Give its name.

   Therefore, the relief on this map segment is

   What is the linear distance between these high and low points?

   In terms of slope, state this distance and relief in *feet per mile*:

   State this in terms of *percent grade* along an ideal slope between these two points (relief in feet ÷ distance in feet): %.

   Show your work:

SECTION 3

Continental Glaciation and the Jackson, Michigan, Topographic Quadrangle

tual Tour
ckson, MI

goo.gl/mdpYSQ

Continental glaciers advanced and retreated over North America and Europe, producing many erosional and depositional features. **Figure 25.7** illustrates some of the most common depositional features associated with the passage of a continental glacier. The unsorted and unstratified deposits of gravel, sand, and clay form moraines, including ground and terminal moraines. Many relatively flat plains of *unsorted* coarse till are formed behind terminal moraines. These **till plains** typically have low, rolling relief, deranged drainage patterns, and the following depositional features. **Drumlins**, smooth hills made of till shaped by the ice, are oriented in the direction of the glacier's movement. **Eskers** are curving, narrow deposits of coarse gravel left by meltwater stream deposits in tunnels beneath the ice. Often eskers end in a delta. **Kames** are small hills of poorly sorted sand and gravel that collected in depressions in the surface of a glacier.

Beyond the morainal deposits, glacio-fluvial **outwash plains** of *stratified drift* feature stream channels that are meltwater-fed, braided, and over-loaded with debris deposited across the landscape. **Kettles** are depressions left by melting blocks of ice that were buried in the drift and are found in both ground moraines and in outwash plains.

▲ Figure 25.7 Common depositional landforms produced by glaciers

Activities and completion items related to continental glaciation:

1. **Figure 25.8** is a hypothetical landscape that has been shaped by continental glaciers. Note the unsorted and unstratified deposits of gravel, sand, and clay that form moraines, including terminal moraines (A) and interlobate moraines (B).

After reviewing Figures 25.7 and 25.8, answer the following questions about Figure 25.8:

If you were looking for drumlins, which letters would you look for?

Which letter shows kames?

If you wanted to walk along the crest of an esker, which letter would you look for?

If you were looking for kettles, which letters would you look for?

Which letter is located on the ground moraine?

Which letter is located on the outwash plain?

CONTOUR INTERVAL 20 FEET

Figure 25.8 Contour map of a continental glaciated landscape.

Once again refer to Figure 24.2g in Lab Exercise 24 and the review of drainage basins and drainage patterns. Imagine the deranged drainage patterns that will result from the disruption of former surface patterns by the erosional and depositional work of continental glaciation.

Keep Figures 25.7 and 25.8 in mind when analyzing Figure 25.9 Topographic Map of Jackson, Michigan, a formerly glaciated area around Jackson, Michigan.

2. Refer to Figure 25.9, the Jackson, Michigan, topographic quadrangle. What is the scale of this map?

   Convert this to a verbal scale. Show your work.

   What is the contour interval?

3. A sinuously curving, narrow ridge of coarse sand and gravel is called an esker. Eskers form along the channel of a meltwater stream that flows beneath a glacier, in an ice tunnel, or between ice walls beneath the glacier. As a glacier retreats, the steep-sided esker is left behind in a pattern roughly parallel to the path of the glacier.

   Locate and identify by name a prominent esker on the map:

   What is the average elevation of this esker?

   How long is this esker, in miles? (Use a string placed along the esker, then pull it straight and compare it to the scale.)

4. Identify the drainage pattern for the portion of the topo map that is generally covered by the Kalamazoo moraine (southwest part of map). Explain what evidence you used to determine this.

5. Use a dark blue colored pencil or pen to trace the drainage of the Grand River. Add arrows to indicate the direction it is flowing.

6. Is this area well drained or poorly drained and swampy? Explain your answer.

▲ Figure 25.9

7. Sometimes an isolated block of ice, perhaps more than a kilometer across, persists in a *ground moraine*, an outwash plain, or valley floor after a glacier retreats. Perhaps 20 to 30 years are required for it to melt. In the interim, material continues to accumulate around the melting ice block. When the block finally melts, it leaves behind a steep-sided hole. Such a feature then frequently fills with water. This is called a kettle. Locate several named kettles that are ponds or lakes on this map and identify their general location:

8. What type of feature is Hatt Hill? Briefly describe how it was formed. Use a red colored pencil or pen and circle several other of these features on the map.

9. If you were walking from Hammer Lake to Blue Ridge, would you be walking over a till plain or an outwash plain? Explain.

10. What is the highest elevation on the map segment? Location and name:

    What is the lowest elevation on the map segment? Location and name:

Mastering Geography

Looking for additional lab prep and review materials? Visit the Study Area in Mastering Geography to enhance your geographic literacy and spatial reasoning skills by accessing MapMaster™ interactive maps, videos, *Mobile Field Trips*, *Project Condor* Quadcopter Videos, *In the News* articles, flashcards, web links, self-study quizzes, and other resources. Contact your professor to find out how you can access Mastering Geography.

Video
Exercise 26
Pre-Lab Video

http://goo.gl/Q2O0B1

Scan to view the
Pre-Lab video

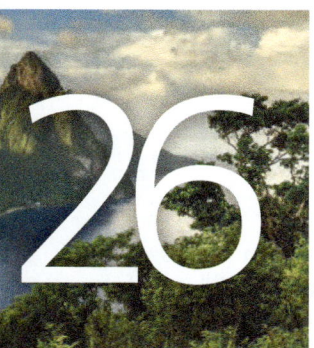

LAB EXERCISE

26 Topographic Analysis: Coastal and Arid Geomorphology and Sea-Level Rise

In addition to streams of water and ice—rivers and glaciers—carving their way across Earth's surface, other exogenic processes are also at work. Waves are relentless in their efforts to reshape the contours of the continents, whether sandy **beaches** or rocky cliffs, along the narrow contact zone between the landmasses and oceans. The coast is also a contact zone between anthropogenic (human-caused) processes and natural processes, including sea-level rise. And landforms that are characteristic of arid climates are limited to global areas where adequate precipitation is always lacking, allowing wind and intermittent streams to remodel the surface.

Topographic map interpretation continues in this lab exercise. Specific geomorphic processes and the resulting landforms from coastal (wave) action and the work of streams in arid climates are analyzed using topographic maps, aerial imagery, and illustrations. Once again, refer to the complete legend of topographic map symbols located on the inside front cover of this lab manual. Lab Exercise 26 features mobile-friendly QR links to Google Earth virtual tours that fly you to the topographic map location, as well as optional online Google Earth activities with downloadable media and worksheets at mygeoscienceplace.com and the Mastering Geography Study Area. Note: The topographic maps and aerial imagery are numbered so that you can compare the topographic map side by side with the aerial imagery.

Key Terms and Concepts

alluvial fan	beach	littoral zone
bajada	graben	playa
barrier spit	horst	sea cliff
basin and range	interior drainage	tombolo
bay barrier	lagoon	wave-cut platform

KEY**LEARNING**concepts

After completion of this lab, you should be able to:

- *Analyze* and *describe* geomorphic processes using topographic maps and Google Earth™ imagery.
- *Interpret* and *describe* several coastal processes and the characteristic landscapes that result, including erosional and depositional features.
- *Analyze* sea level rise projections and *identify* potential areas of inundation.
- *Interpret* and *describe* several arid climate processes and the characteristic landscapes that are produced, including alluvial fans, buttes, and mesas.

Materials/Sources Needed

pencil, colored pencils, or pens	calculator
ruler	Google Earth

Lab Exercise and Activities

SECTION 1

Coastal Features and the Duncans Mills and Arched Rock Topographic Map and Aerial Imagery

Most of Earth's coastlines are relatively new and are settings for continuous change. The land, ocean, atmosphere, Sun, and Moon interact to produce tides, currents, and waves that create erosional and depositional features along the continental margins. A dynamic equilibrium among the energy of waves, wind, and currents; the supply of materials; the slope of the coastal terrain; and the fluctuation of relative sea level produces coastline features of infinite variety. The coastal environment is called the **littoral zone.** (Littoral comes from the Latin word *litus*, meaning "seashore.") The littoral zone spans both land and water. Landward, it extends to the highest water line that occurs on shore during a storm. Seaward, it extends to the point at which storm waves can no longer move sediments on the seafloor (usually at depths of approximately 60 m, or 200 ft). The specific contact line between the sea and the land is the *shoreline*, and adjacent land is considered the *coast*.

The active margins of the Pacific along the North and South American continents are characteristic coastlines affected by erosional landform processes. Erosional coastlines tend to be rugged, of high relief, and tectonically active, as expected from their association with the leading edge of drifting lithospheric plates (see the plate tectonics discussion in Lab Exercise 19).

Figure 26.1 presents features commonly observed along an erosional coast. **Sea cliffs** are formed by the undercutting action of the sea. As indentations are produced at water level, such a cliff becomes notched, leading to subsequent collapse and retreat of the cliff. A **wave-cut platform**, or *wave-cut terrace*, is formed by wave action as a horizontal bench in the tidal zone.

If the relationship between the land and sea level has changed over time, multiple platforms or terraces may arise like stairsteps back from the coast. Other erosional forms evolve along cliff-dominated coastlines, including *sea caves, sea arches*, and *sea stacks*. As erosion continues, arches may collapse, leaving isolated stacks in the water.

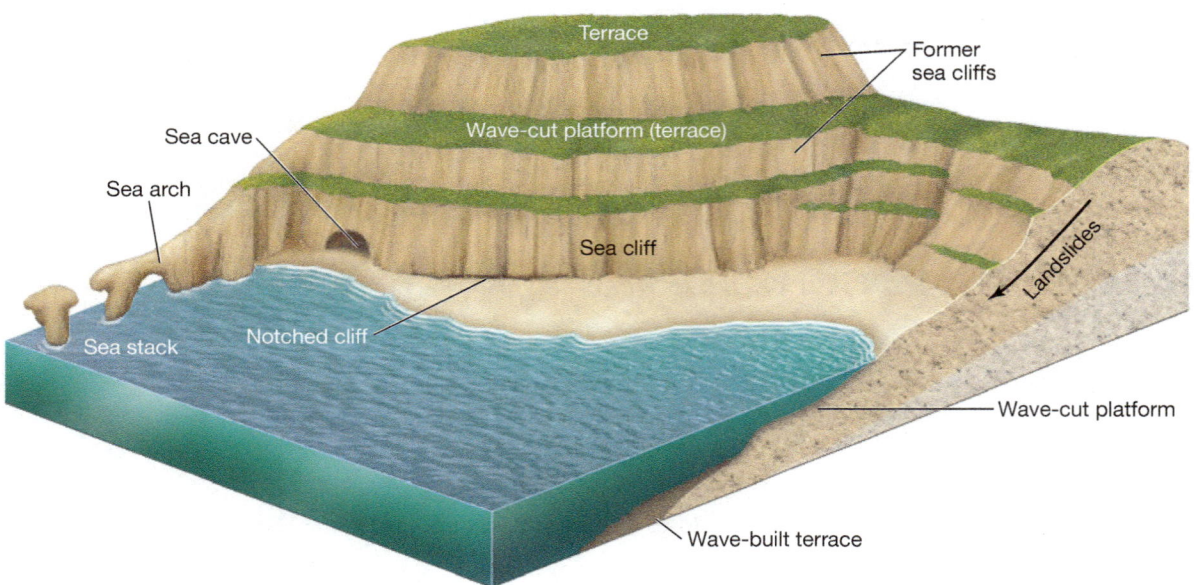

▲ **Figure 26.1** Characteristic coastal erosional landforms

Depositional coasts generally are located near onshore plains of gentle relief, where sediments are available from many sources, although such features can occur along all coasts. Characteristics of wave- and current-deposited landforms are illustrated in **Figure 26.2** and may involve sediments of varying sizes. A **barrier spit** consists of material deposited in a long ridge, attached at one end, extending out from a coast; it partially blocks the mouth of a bay, except for a *tidal inlet*. A spit becomes a **bay barrier**, sometimes referred to as a *baymouth bar*, if it completely cuts off the bay from the ocean and forms an inland **lagoon**. Tidal flats and salt marshes are characteristic low-relief features wherever tidal influence is greater than wave action.

A **tombolo** occurs when sediment deposits connect the shoreline with an offshore island or sea stack. A tombolo forms when sediments accumulate on a wave-built terrace that extends below the water.

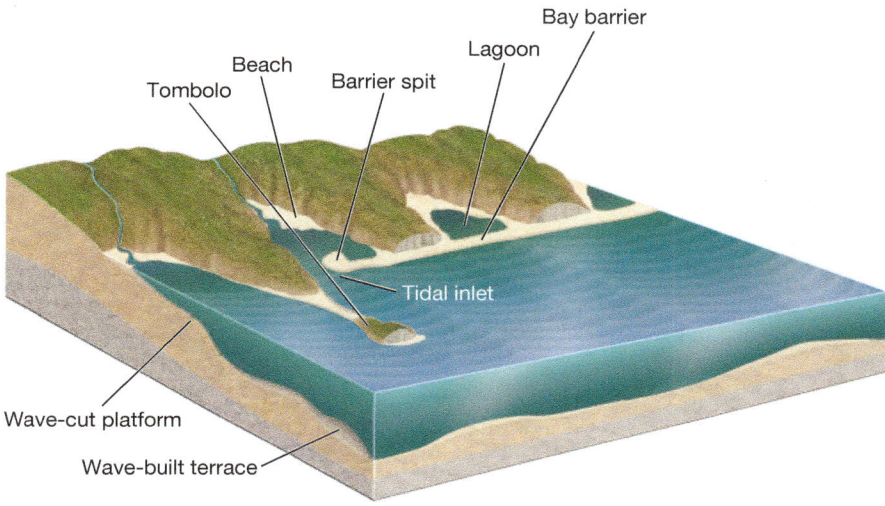

▲ **Figure 26.2** Characteristic coastal depositional landforms

Activities and completion items related to coastal geomorphology:

Virtual Tour
Duncans Mills and Arched Rock, CA

https://goo.gl/BMQ3gX

1. Refer to **Figure 26.6** Topographic Map of Duncans Mills and Arched Rock, CA and **Figure 26.8** Aerial Imagery of Duncans Mills and Arched Rock, CA.

   a) What is the scale of this map? _____

   b) What is the contour interval? _____

2. What type of feature is Goat Rock? is it an erosional or depositional feature? What type of feature would it be if it wasn't connected to the mainland by sediment?

3. What type of feature is found between Jenner Beach and Goat Rock? Describe how it appears and how it differs from other similar features.

4. List and name three erosional and depositional features found in Figure 26.6. Circle and label them on Figure 26.6.

5. If you were in a small boat off of Jenner Beach and wanted to come ashore, would you come ashore on the north or south end of Jenner Beach? Why? Using the topographic map symbol key on the inside front cover of your lab manual, what are the features offshore from Jenner Beach?

6. Marine terraces are remarkable indicators of an emerging coastline. What evidence do you find of such a leveled terrace on the map? Describe where the terraces are and at what elevation.

7. Circle and label some wave-cut cliffs along the coast on Figure 26.6. Using the contour lines, how high above sea level are they north of Russian Gulch? . . . south of Goat Rock?

8. Describe some specific differences you see in comparing the map and the photos. Are there any cultural features that appear on the photos and not on the map? Describe.

9. Optional Google Earth activity, Duncans Mills, California. For the KMZ file and questions, go to mygeoscienceplace.com or the Mastering Geography Study Area. Then click on the cover of *Applied Physical Geography: Geosystems in the Laboratory*.

SECTION 2

Sea Level Rise in Miami, FL

The effects of rising sea level from climate change are readily observable along U.S. coastlines. As high tides increase, nuisance flooding, also known as "sunny-day flooding," now occurs regularly along the Atlantic coastline, inundating roads, swamping basements, and disrupting storm drainage systems. The real estate web company Zillow estimated in 2016 that almost $1 billion of real estate would be submerged with a six foot rise. In Miami Beach, where nuisance flooding increased 400% from 2006 to 2016, city officials have implemented programs to install storm water pumps and raise road elevations, at a cost of $400 to $500 million. Sea level rise could affect up to 300,000 Miami, FL residents by 2100. Because of it's topography

Florida is the state most at risk from sea level rise. Nineteen of the 30 cities with the largest populations of people potentially affected by sea-level rise are in Florida.

The rate of rise along the U.S. East Coast is greater than the rate along the U.S. West Coast. Sea level is now rising more quickly than predicted by most climate models, and the rate appears to be accelerating. During the last century, sea level rose 6.7–8.3 in. (17–21 cm), a greater rise in some areas (such as the U.S. Atlantic coast) than at any time during the past 3000 years. From 1901 to 2010, tidal gauge records show that sea level rose 0.07 in. (1.7 mm) per year. From 1993 to 2013, satellite data show that sea level rose 0.13 in.(3.4 mm) per year.

The tide station on Virginia Key, just south of Miami, has recorded an increase of 0.22" (5.6 mm) per year for the time period of 1994 to 2016.

NOAA scientists developed scenarios for sea-level rise based on present ice-sheet losses coupled with losses from mountain glaciers and ice caps worldwide. These models project a rise of 0.3 m (1.0 ft) at the low end of the range, under the lowest emission scenarios. On the high end, 1.2 m (4.0 ft) is plausible, with up to 2.0 m (6.6 ft) as the extreme. In the United States, sea level is rising more quickly on the East Coast than on the West Coast. The largest areas of inundation at a rise of only 1 m are around the low-elevation cities of Miami and New Orleans. Many states, including California, presently use a 4.6 ft. (1.4 m) sea-level rise this century as a standard for planning purposes. The effects of sea-level rise will vary from place to place, with some world regions more vulnerable than others. Even a small sea-level rise would bring higher water levels, higher tides, and higher storm surges to many regions, particularly impacting river deltas and low-lying mainland areas. Among the densely populated cities most at risk are New York, Miami, and New Orleans in the United States, Mumbai and Kolkata in India, and Shanghai in China, where large numbers of people will have to leave coastal areas. A 1.0 ft (0.3 m) rise would produce a shoreline retreat of 98 ft (30 m) in some locations, while a 3.2 ft (1.0 m) rise would displace an estimated 130 million people, and a 6.6 ft (2.0 m) rise would displace 500 million people. The social and economic consequences of rising sea level especially affect small, low elevation island states. A flood of environmental refugees driven by climate change would be expected to continue for decades. Sea level increases will continue beyond 2100, even if greenhouse gas concentrations were to be stabilized today.

1. Compare the rate of sea-level rise at Virginia Key in Florida from 1994–2016 with the average rate from 1993-2013. What was increase per year at Virginia Key? What was the average rate?

2. If sea level at Virginia Key continued to rise at that same rate from 2016 on, how much higher would sea level be in 2100 at Virginia Key? How does that rise compare with the estimated intermediate and extreme projections shown in **Table 26.1**? Based on these answers, is the rate of sea-level rise increasing or decreasing?

One of the highest storm surges recorded along the east coast was during the Great Miami Hurricane of 1926, a category 4 hurricane with sustained winds of 150 mph. 11.7 feet (3.6 m) of storm surge was recorded along Biscayne Boulevard in downtown Miami. We'll assume another storm of this magnitude for this exercise.

Table 26.1 shows projected sea level rise data for Miami, Fl. **Figure 26.7** shows current elevation above mean high water for the Miami Beach area. That is, the elevations shown represent the elevation above the average high tide, not above mean sea level. **Figure 26.9** is a topographic map of the same region.

3. Use a dark blue colored pencil or pen to draw the approximate locations of the following on Figure 26.7: the region that will be inundated under the extreme projection by 2040, 2060, and 2100. Use a red pencil or pen to draw the approximate location of the boundary of the region that could be inundated by extreme sea level rise and storm surge in 2100.

TABLE 26.1 Projected sea level rise for Miami, Fl

Year	Intermediate	Extreme	Extreme Storm Surge	Extreme Sea Level Rise + Extreme Storm Surge
Current			11.7 ft (3.6 m)	
2040	1.05 ft (0.3 m)	2.00 ft (0.6 m)	11.7 ft (3.6 m)	13.7 ft (4.2 m)
2060	1.8 ft (0.5 m)	4.00 ft (1.2 m)	11.7 ft (3.6 m)	15.7 ft (4.8 m)
2100	3.8 ft (1.1 m)	10.4 ft (3.2 m)	11.7 ft (3.6 m)	22.1 ft (6.7 m)

4. Circle the following locations on Figure 26.9 and list their current elevation above sea level and the depth of water at that location under the extreme scenario:

Name	Current elevation	Water depth in 2100
a) The Miami Seaquarium on Virginia Key	_____	_____
b) Dodge Island (The Port of Miami)	_____	_____
c) The water treatment plant on the north end of Virginia Key	_____	_____
d) The islands along Venetian Way between Miami and Miami Beach	_____	_____
e) The city cemetery	_____	_____

5. How far inland in the city of Miami would you go to find a location with an elevation higher than 15′?

6. Can you find any locations in Miami north of the Miami River, east of Highway 95, and south of Highway 195, with an elevation that is higher than the projected extreme sea level rise in 2100? extreme sea level rise with storm surge in 2100?

7. Can you find any locations in Miami Beach with an elevation that is higher than the projected extreme sea level rise in 2100? extreme sea level rise with storm surge in 2100?

8. Will the effects of sea level rise be greater in Miami or Miami Beach? Explain your answer and provide examples.

9. What do you think the future holds for the real estate market in Miami?

SECTION 3

Death Valley, California

Condor
Characteristics of Alluvial Fans

https://goo.gl/YjqXQT

Mobile Field Trip
Desert Geomorphology

https://goo.gl/muobfN

Dry climates occupy about 26% of Earth's land surface and, if all semiarid climates are considered, perhaps as much as 35% of all land, constituting the largest single climatic region on Earth. Despite the general dryness, water remains the major erosional force in arid and semiarid regions.

In arid climates, with their intermittent water flow, a particularly noticeable fluvial landform is the **alluvial fan**, or *alluvial cone*, which occurs at the mouth of a canyon where it exits into a valley (**Figure 26.3**). The fan is produced by flowing water that loses velocity as it leaves the constricted channel of the canyon and, therefore, drops layer upon layer of sediment along the base of the mountain block. Water then flows over the surface of the fan and produces a braided drainage pattern, shifting from channel to channel with each flood event.

An interesting aspect of an alluvial fan is the natural sorting of materials by size. Near the mouth of the canyon, boulders and gravels are deposited, grading slowly to pebbles and finer gravels with distance out from the mouth. Then sands and silts are

▲ **Figure 26.3** Typical arid landscape features.

deposited, with the finest clays and salts carried in suspension and solution all the way to the valley floor. In the wet season, a shallow lake may form that will evaporate and form a dry lake bed, or **playa**. Over time, as more sediment is deposited by the shifting stream channels, the alluvial fans will begin to coalesce into a continuous surface called a **bajada**. **Interior drainage** occurs when streams do not flow to the ocean but instead end in inland basins or seas. Playas are an indication of interior drainage.

Earth's crust is being stretched by tectonic activity, forming the **basin and range** topography of the American west. The stress of tension forms normal faults, which creates a landscape of mountain ridges called **horsts** and down-dropped fault blocks called **grabens**. Death Valley, California, is one such graben formed by the stretching of Earth's crust. Tectonic forces have created a juxtaposition of the lowest spot in North America, −86 m (−282 ft.) at Badwater in Death Valley, only 240 km (150 mi.) from the highest point in the contiguous lower 48 United States, Mt. Whitney, at 4421m (14,421 ft.).

Examples of these features are found in Death Valley, California, presented as Topographic Map #9.

Analysis and completion items for **Figure 26.10**, Topographic Map of Death Valley, CA, and **Figure 26.12**, Aerial Imagery of Death Valley, CA:

1. This topographic map has two different contour intervals. What is the contour interval for elevations below 1000 ft.? What is the contour interval on this map for elevations over 1000 ft.? How would the upper elevations look if they had been mapped with the same contour interval as the lower elevations? How would the lower elevations look if they had been mapped at the same contour interval as the higher elevations?

Virtual Tour
Death Valley

https://goo.gl/s08ua5

What is the scale of this map?

2. Although Death Valley receives less than 5 cm (2 in.) of rain each year, water is an important agent in modifying the landscape. List by name at least three water features formed by running or standing water (not springs) and trace or outline them on the topographic map with a blue colored pencil or pen.

3. Using a black colored pencil or pen, label the following features on **Figure 26.10**: bajada, alluvial fan (may be part of a bajada), playa, and graben.

4. Alluvial fans are important sources of water in arid climates. Rain from storms percolates through the alluvium of the fan before reaching an impermeable layer. Some of the signs of this are springs and vegetation at the bottom of alluvial fans. Identify by name at least four locations on Figure 26.10 with springs or vegetation and circle them on the topographic map with a blue colored pencil or pen.

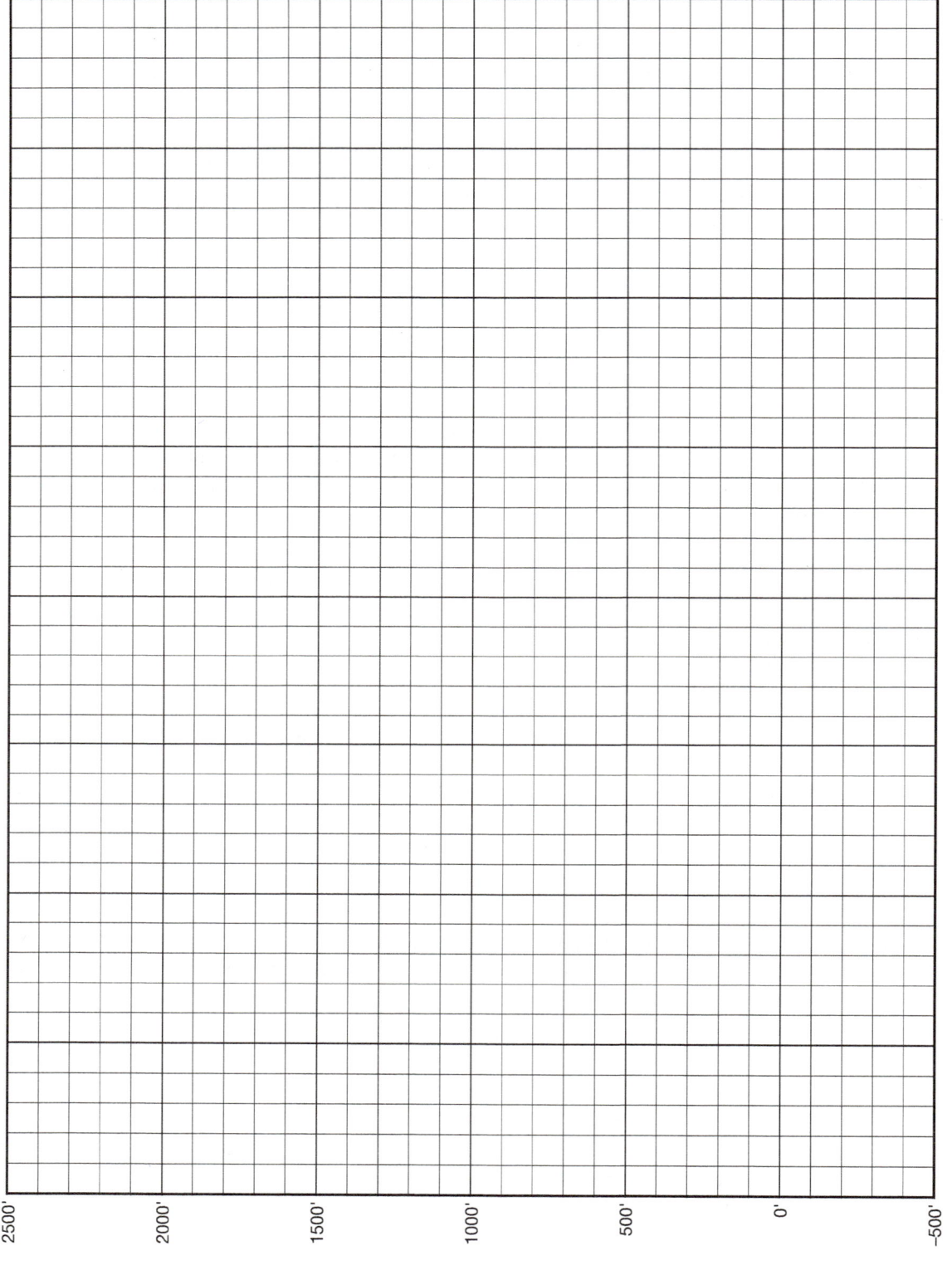

2500'

2000'

1500'

1000'

500'

0'

−500'

▲ **Figure 26.4** Topographic profile across Death Valley, CA

5. Use **Figure 26.4** to plot the topographic profile along the 36° 20′ N parallel. The 36° 20′ N parallel is not marked on the map, but there is a tick mark on the map at 36° 20′ N on the west side of the map. Start your profile at the 36° 20′ N tick mark just north of Trail Canyon on the west and proceed east across the valley to the 800-ft elevation contour on Figure 26.9, Topographic Map of Death Valley, CA.

6. What is the linear distance from the 1400-ft elevation contour in Trail Canyon to the 200-ft contour? What is the relief between these two points?

    In terms of slope, state this distance and relief in feet per mile:

    State this in terms of percent grade:

7. Using a black colored pencil or pen, trace the outline of the alluvial fan at the mouth of Trail Canyon. What is the lowest-value elevation contour line on the fan? Using a blue colored pencil or pen, trace the streams on the Trail Canyon alluvial fan.

8. Briefly describe how alluvial fans are formed. Why do you think there is more than one stream on the Trail Canyon alluvial fan? Relate your answer to the process by which alluvial fans are formed.

9. Explain why you think the alluvial fans on the west side of Death Valley are larger than the alluvial fans on the east side of Death Valley.

10. Identify two features that are more obvious in Figure 26.10 Topographic Map of Death Valley, CA, than in Figure 26.12 Aerial Imagery of Death Valley, CA, and two figures that are more obvious in the aerial imagery than in the topographic map.

11. Optional Google Earth™ activity, Cedar Creek Alluvial Fan in Ennis, Montana. For the KMZ file and questions, go to mygeoscienceplace.com or the Mastering Geography Study Area. Then click on the cover of *Applied Physical Geography: Geosystems in the Laboratory.*

SECTION 4

Mitten Buttes, Arizona, Quadrangle and Aerial Imagery

Monument Valley straddles the Utah–Arizona border west of the Four Corners area. It is almost entirely within the Navajo Indian Reservation and is accessed through the Navajo Tribal Park off U.S. 163. Part of this region is shown in **Figure 26.11**, Topographic Map of Mitten Buttes, AZ, and in **Figure 26.13** Aerial Imagery of Mitten Buttes, AZ.

The mesas, buttes, and pinnacles (listed in decreasing size and increasing age) of arid landscapes are resistant horizontal rock strata that have eroded differentially. Removal of the less-resistant sandstone strata produces unusual desert sculptures—arches, windows, pedestals, and delicately balanced rocks. Specifically, the upper layers of sandstone along the top of an arch or butte are more resistant to weathering and protect the sandstone rock beneath.

The removal of all surrounding rock through differential weathering leaves enormous buttes as residuals on the landscape. If you imagine a line intersecting the tops of the Mitten Buttes shown in **Figure 26.5**, you can gain some idea of the quantity of material that has been removed. These buttes exceed 300 m (1000 ft) in height, similar to the Chrysler Building in New York City or First Canadian Place in Toronto.

Examine Figure 26.11, Topographic Map of Mitten Buttes, AZ, and Figure 26.13, Aerial Imagery of Mitten Buttes, AZ, to answer the following:

1. What is the contour interval on this map segment?

   What is the scale of this map?

2. What is the elevation of the Monument Valley Visitor Center? What is the relief between the top of Mitchell Mesa on the west side of the map and the lower elevations along the east side of the map?

3. Which is the tallest named butte? What is its elevation?

▲ **Figure 26.5** Rock removal by weathering, erosion, and transport in Monument Valley

4. Examining these features on the topo map and in the aerial imagery, can you describe any features that indicate differing rock resistances to weathering (slopes = less resistance; cliffs = more resistance)? Describe and explain.

5. Use a dark blue pencil or pen to draw an arrow in the direction of flow of West Gypsum Creek. Which type of stream is West Gypsum Creek?

6. Earlier we stated that water is the major erosional force in the desert; although occurring infrequently, when it rains, the downpour is usually intense, and floods flash across the sparsely vegetated surfaces swiftly. Given this reality, and using the map and aerial imagery, describe several examples that demonstrate to you that water is the major erosional force and is occurring infrequently. List at least four examples (look for dry washes, sharp features, and alluvial deposits). Name at least one type of feature that is missing from the Mitten Buttes map that is present on the Death Valley map.

7. Optional Google Earth™ activity, Mitten Buttes, Arizona. For the KMZ file and questions, go to mygeoscienceplace.com or the Mastering Geography Study Area. Then click on the cover of *Applied Physical Geography: Geosystems in the Laboratory*.

Mastering Geography

Looking for additional lab prep and review materials? Visit the Study Area in Mastering Geography to enhance your geographic literacy and spatial reasoning skills by accessing MapMaster™ interactive maps, videos, *Mobile Field Trips*, *Project Condor* Quadcopter Videos, *In the News* articles, flashcards, web links, self-study quizzes, and other resources. Contact your professor to find out how you can access Mastering Geography.

▲ **Figure 26.6** Topographic Map of Duncans Mills and Arched Rock, CA

West Branch
Russian Gulch

Middle Branch
Russian Gulch

East Branch
Russian Gulch

123°09'

123°08'

123

Russian Gulch

— 38°28'

— 38°27'

Jenner
Beach

Crab
Island

Penny Island

Goat
Rock
Beach

Halfmile
Rock

Goat
Rock

Mile Rocks

Blind
Beach

Arched
Rock

WILLIG DR

CALIFORNIA

QUADRANGLE LOCATION

— 38°26'

123°09'

123°08'

GOAT

Pea
Hill

| 1 | 0.5 | 0 | 1 |
| MILES |

| 1000 | 0 | 1000 | 2000 | 3000 | 4000 | 5000 | 6000 | 7000 |
| FEET |

| 1 | 0.5 | 0 | 1 |
| KILOMETERS |

CONTOUR INTERVAL 40 FEET   SCALE 1: 24,000

38

38

123°

Gull Rock

▲ **Figure 26.7** Elevation Above Mean High Water, Miami and Miami Beach

Elevation Above Mean High Water

- ≤ 3'
- 3–6'
- 6–9'
- 9–12'
- 12–15'
- 15–18'
- 18–21'
- 21–24'
- > 24'

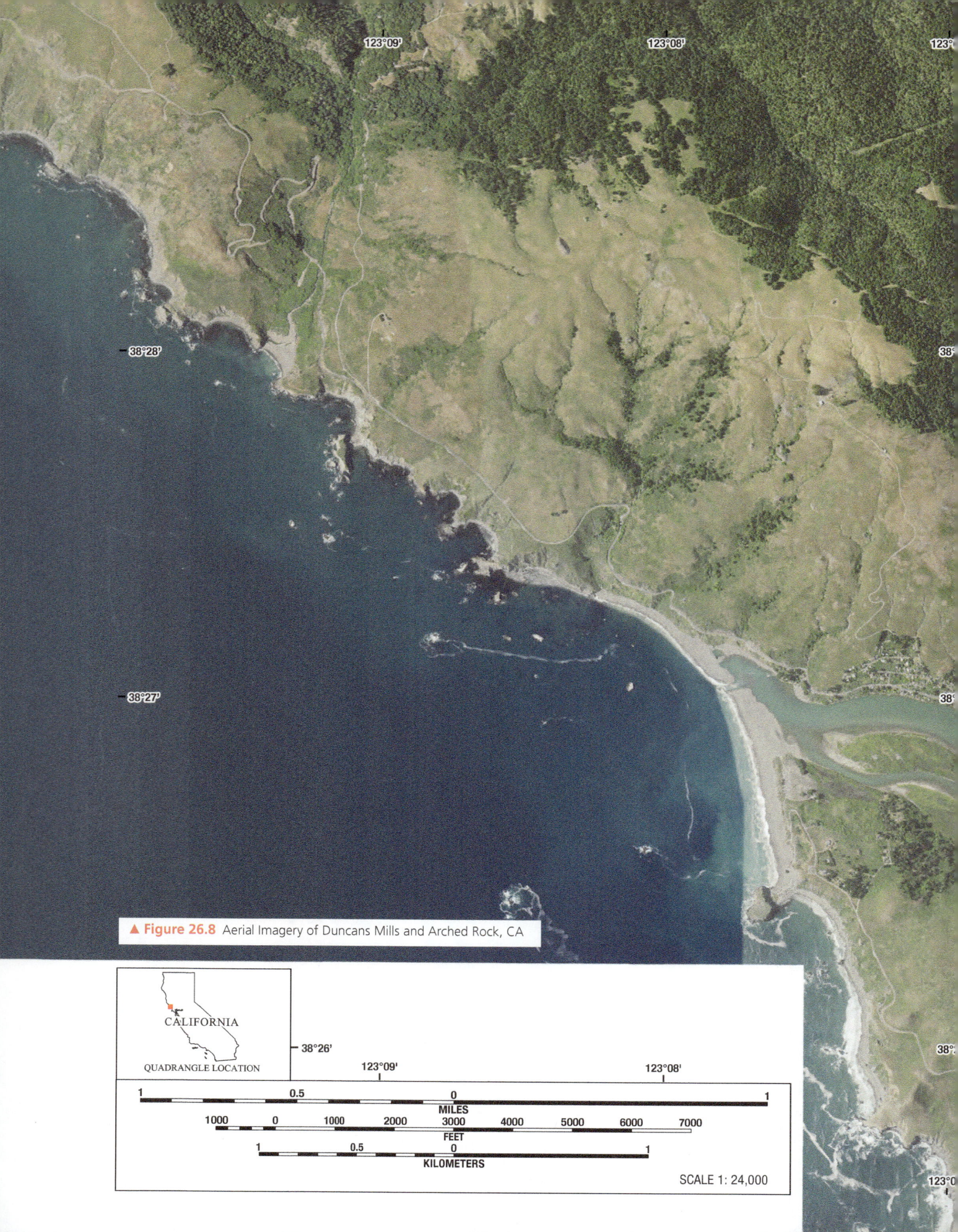

▲ **Figure 26.8** Aerial Imagery of Duncans Mills and Arched Rock, CA

123°09'   123°08'   123°

38°28'

38°27'

CALIFORNIA

QUADRANGLE LOCATION

38°26'

123°09'   123°08'   123°

38°

38°

38°

123°0

MILES

1   0.5   0   1

1000   0   1000   2000   3000   4000   5000   6000   7000

FEET

1   0.5   0   1

KILOMETERS

SCALE 1: 24,000

▲ **Figure 26.9** Topographic Map of Miami and Miami Beach

▲ **Figure 26.10** Topographic Map of Death Valley, CA

SCALE 1: 150,000

▲ **Figure 26.11** Topographic Map of Mitten Butte, AZ

▲ **Figure 26.12** Aerial Imagery of Death Valley, CA

CALIFORNIA

LOCATION

4	2	0	4
MILES			

4	2	0	4
KILOMETERS			

SCALE 1: 150,000

110°06' 110°05' 110°04'

36°59' 36°59'

36°58' 36°58'

36°57' 36°58'

MITTEN BUTES, AZ

ARIZONA

QUADRANGLE LOCATION

▲ **Figure 26.13** Aerial Imagery of Mitten Butte, AZ

110°06' 110°05' 110°04'

NOTES

Name: _____   Laboratory Section: _____

Date: _____   Score/Grade: _____

LAB EXERCISE

27

Topographic Analysis: Karst Landscapes

Limestone is abundant on Earth, and many landscapes are composed of it. These areas are quite susceptible to chemical weathering. Such weathering creates a specific landscape of pitted- and bumpy-surfaced topography, limited surface streams due to infiltration, and well-developed solution channels (dissolved openings and conduits) underground. Remarkable labyrinths of underworld caverns also may develop due to weathering and erosion caused by groundwater.

These are the hallmarks of **karst topography**, named for the Krš Plateau in Slovenia, where karst processes were first studied. Approximately 15% of Earth's land area has some karst features, with outstanding examples found in southern China, Japan, Puerto Rico, Cuba, the Yucatán of Mexico, Kentucky, Indiana, New Mexico, and Florida. As an example, approximately 38% of Kentucky has **sinkholes** and related karst features noted on topographic maps.

For a limestone landscape to develop into karst topography, there are several necessary conditions:

- The limestone formation must contain 80% or more calcium carbonate for solution processes to proceed effectively.

- Complex patterns of joints in the limestone are needed for water to form routes to subsurface drainage channels.
- There must be an air-filled zone between the ground surface and the water table.
- Vegetation cover supplies varying amounts of organic acids and carbon dioxide that enhance the solution process.

The role of climate in providing optimum conditions for karst processes remains in debate, although the amount and distribution of rainfall appears important. Karst occurs in arid regions, but here it was primarily formed in prior times with greater humidity and precipitation.

Areas of karst may exhibit diversion of surface water to underground flows. A stream may appear and disappear along its channel, called a **disappearing stream**. Groundwater and surface waters dynamically interact through the jointed rock structures and permeable rock. Lab Exercise 27 features mobile-friendly QR links to Google Earth virtual tours that fly you to the topographic map location, as well as optional online Google Earth activities with downloadable media and worksheets at mygeoscienceplace.com and the Mastering Geography Study Area.

Key Terms and Concepts

disappearing stream   karst topography   sinkhole

KEY**LEARNING**concepts

After completion of this lab, you should be able to:

- *Analyze* and *describe* karst landscapes by using topographic maps and aerial imagery.
- *Relate* these surface features to groundwater and the underground world of solution caverns.

Materials/Sources Needed

pencil   ruler
calculator   Google Earth

Lab Exercise and Activities

SECTION 1

An idealized karst landscape with karst topography

▲ **Figure 27.1** Idealized features of a karst landscape

Virtual Tour
Oolitic, IN

https://goo.gl/0cq5YQ

Oolitic, Indiana, Quadrangle and Aerial Imagery

The weathering of limestone landscapes creates many sinkholes, which form in circular depressions. Traditional studies may call a sinkhole a *doline*. Water from the surface can flow through joints or fractures and slowly dissolve the rock around entry, forming a pit called a solution sinkhole. Another type of sinkhole is formed if an underground cavern's roof collapses. Everything above the cavern (rocks, soil, trees, buildings) will drop into the cavern and form a collapse sinkhole. A gently rolling limestone plain might be pock-marked by sinkholes with depths of 2 to 100 m (7 to 330 ft) and diameters of 10 to 1000 m (33 to 3300 ft). Using **Figure 27.2** Topographic Map of Oolitic, Indiana and **Figure 27.4** Aerial Imagery of Oolitic, Indiana, answer the following. Note: The topographic map and aerial imagery figure pairs are printed out of order so that you can compare Figures 27.2 and 27.4 side by side, and compare **Figures 27.3** and **27.5** side by side.

1. Refer to Figure 27.2 Topographic Map of Oolitic, Indiana. What is the scale of this map?

   What is the contour interval?

2. In the space provided draw the topographic map symbol that indicates a sinkhole; sketch a contour line and illustrate this symbol. (Check out the chart of topographic map symbols inside the front cover of this manual.)

Find Section 31 on the map (west of the highway, west center of map). Section 31 is marked on the north and east by a red section line, on the south by a black-dashed township line, and on the west by 500 West Road.

Answer the following about Section 31:

3. What are the highest and lowest elevations in Section 31?

   Highest:

   Lowest:

4. How many sinkholes are there in the 1 mi^2 section?

   Do any of these sinkholes have water in them (implying that they must have impermeable soil or rock along their bases)?

   Examine Figure 27.4, Aerial Imagery of Oolitic, Indiana.

5. Describe any economic activity specific to this karst environment that you see on the topo map and aerial imagery.

6. Describe how the sinkholes appear in the photos.

7. Can you distinguish between older and newer mining areas by using the aerial imagery? How were you able to determine this observation? Discuss.

8. This limestone region is deeply dissected by Salt Creek, flowing along the east side of the map and photos. The entrenched meanders east of Oolitic are remarkable in their geometric (rock-structure controlled) bends. Such an entrenched stream is important in the development of a karst region because it permits a continuous movement of water through the jointed limestone landscape. What is the local relief between the town and the creek? Examine the contour lines carefully—they are tightly spaced along the cliffs.

9. Optional Google Earth™ activity, Oolitic, Indiana. For the KMZ file and questions, go to mygeoscienceplace.com or the Mastering Geography Study Area. Then click on the cover of *Applied Physical Geography: Geosystems in the Laboratory.*

SECTION 2

Lake Wales, Florida, Quadrangle and Aerial Imagery

Virtual Tour
Lake Wales, FL

https://goo.gl/FioVQV

In Florida, several sinkholes have made news because water tables lowered by groundwater pumping have caused their collapse into underground solution caves, taking with them homes, businesses, and even new cars from an auto dealership. One such sinkhole collapsed in a suburban area in 1981 and others in 1993 and 1998. The area around Lake Wales, Florida (south of Orlando and north of the Everglades), is characteristic of limestone with karst topography, high water tables, some dry depressions, limited surface streams, marsh lands, and groves of citrus trees—see Figure 27.3, Topographic Map of Lake Wales, Florida, and Figure 27.5, Aerial Imagery of Lake Wales, Florida. Complete the following items relative to this map.

1. Refer to Figure 27.3 Topographic Map of Lake Wales, Florida. What is the scale of this map?

   What is the contour interval?

2. How many sinkhole depressions contain lakes in the map segment?

3. Do any of these lakes sit in a depression (give a count) 10 ft deep?

   15 ft deep?

   20 ft deep?

   More than 25 ft deep?

4. Water will flow downslope between sinkholes. Using lake- and pond-surface elevation as an indicator, in which direction do you think groundwater flows through this area? Explain your thinking. Using a dark blue colored pen, draw arrows to indicate the direction of flow of groundwater.

5. A hospital, the Lake Wales Medical Center, is shown south (SSE) of Lake Wales. Determine the elevation of the hospital as well as the elevation of the lake surface of Lake Wales.

   Hospital: Lake Wales:

   If you were standing at the hospital, could you see the lake? Explain.

6. Evaluate the status of the elevation of the water table over time on this map. Is the water table rising or falling? Using a dark green colored pencil or pen, circle at least three features on the map that support your statement. Describe why these three features support your assertion.

7. Do you see any evidence that the water table is at the surface or that the region is poorly drained? Describe at least two types of features that would show a high water table or poor drainage.

8. What is the primary activity on the landscape? Take a look at Sections 25 and 36. Use the topographic map symbol key inside the front cover of this manual to help identify the activity.

9. Do you find any surface streams on the topographic map? Explain your observation. What would produce this hydrologic situation?

10. List and describe three differences you can see between the topographic map and the aerial imagery. How does each one appear differently on the map and on the aerial imagery.

11. Optional Google Earth™ activity, Lake Wales, Florida. For the KMZ file and questions, go to mygeoscienceplace.com or the Mastering Geography Study Area. Then click on the cover of *Applied Physical Geography: Geosystems in the Laboratory*.

Mastering Geography

Looking for additional lab prep and review materials? Visit the Study Area in Mastering Geography to enhance your geographic literacy and spatial reasoning skills by accessing MapMaster™ interactive maps, videos, *Mobile Field Trips, Project Condor* Quadcopter Videos, *In the News* articles, flashcards, web links, self-study quizzes, and other resources. Contact your professor to find out how you can access Mastering Geography.

▲ Figure 27.2

▲ Figure 27.3

86°34' 86°33' 86°32'

38°56'   38°56'

38°55'   38°55'

38°54'   38°54'

86°34'   86°33'   86°32'

38°53'

▲ Figure 27.4

▲ Figure 27.5

NOTES

Name: _____   Laboratory Section: _____

Date: _____   Score/Grade: _____

LAB EXERCISE

28 Soils

Earth's landscape generally is covered with soil. **Soil** is a dynamic natural material composed of fine particles in which plants grow, and it contains both mineral fragments and organic matter. The soil system includes human interactions and supports all human, other animal, and plant life. If you have ever planted a garden, tended a house plant, or been concerned about famine and soil loss, this lab exercise will interest you. You may discover that you have made some casual observations that can be applied in this exercise.

Soil science is interdisciplinary, involving physics, chemistry, biology, mineralogy, hydrology, taxonomy, climatology, and cartography. Physical geographers are interested in the spatial patterns formed by soil types and the environmental factors that interact to produce them. **Pedology** concerns the origin, classification, distribution, and description of soil (from the Greek *pedon*, meaning "soil" or "earth"). Pedology is at the center of learning about soil as a natural body, but it does not dwell on its practical uses. *Edaphology* (from the Greek *edaphos*, meaning "soil" or "ground") focuses on soil as a medium for sustaining higher plants. Edaphology emphasizes plant growth, fertility, and the differences in productivity among soils. Pedology gives us a general understanding of soils and their classification, whereas edaphology reflects society's concern for food and fiber production and the management of soils to increase fertility and reduce soil losses.

Observing soils firsthand in the field is the best way to learn about their properties: Construction sites, excavations on your campus or nearby, and roadcuts along highways all provide opportunities for seeing soils in their natural surroundings. In many locales, an *agricultural extension service* can provide specific information and perform a detailed analysis of local soils. Soil surveys and local soil maps are available for most counties in the United States and for the Canadian provinces. Your local phone book may list the U.S. Department of Agriculture, Natural Resources Conservation Service (http://www.nrcs.usda.gov/), or Agriculture Canada's Soil Information System (http://sis.agr.gc.ca/cansis/intro.html).

This exercise gives you an opportunity for some hands-on experience with soils and for using some of the tools and methods that soil scientists use in their work. Lab Exercise 28 features six sections. (*Note: The results for many activities in this exercise will vary, depending on the samples, local sites, and other materials available to your lab.*)

Key Terms and Concepts

clay	porosity	soil horizon
humus	regolith	soil pH (acidity–alkalinity)
loam	sand	soil profile
pedology	silt	soil properties
pedon	soil	soil science
peds	soil classification	soil structure
permeability	soil color	soil texture
polypedon	soil consistence	solum

KEY**LEARNING**concepts

After completion of this lab, you should be able to:

- *Identify* basic components of soil and soil properties.
- *Determine* main components of soil samples by color.
- *Identify* major soil texture categories and *classify* soils by texture.
- *Use* the "feel method" and *determine* texture of soil samples.
- *Demonstrate* and *observe* how soil texture affects porosity and movement of water through soil.
- *Discern* the characteristics and properties of horizons in soil profiles and *identify* horizons.
- *Measure* pH level in soil samples and *determine* the soil pH (acidity or alkalinity).
- *Identify* specific soil series that are featured as state soils.

Materials/Sources Needed

pencils
colored pencils
ruler
soil samples
Munsell Color Chart (optional, 175 colors)
glass jars or beakers

water
soup can with both ends removed (optional)
pH test equipment (handheld meter or color test kit)
soil samples, or soil sample photographs,
   or Internet site of soil profiles
Internet access (optional)

Lab Exercise and Activities

SECTION 1

Soil Color

Soil properties are the characteristics or traits of soil, some of which include **soil color**, **texture**, **structure**, **consistence**, **porosity**, moisture, and chemistry. We examine a few of these properties, beginning with color.

Soil color is one of the most obvious traits, suggesting composition and chemical makeup in mineral soils. If you look at exposed soil, color may be the most obvious trait. Among the many possible hues are the reds and yellows found in soils of the southeastern United States (high in iron oxides); the blacks of prairie soils in portions of the U.S. grain-growing regions and Ukraine (richly organic); and white-to-pale hues found in soils containing silicates and aluminum oxides. Reduced iron imparts gray and greenish colors. In dry areas, calcium carbonate or other water-soluble salts give the soil a white color.

However, color can be deceptive: Soils of high humus content, organic materials from decomposed plant and animal litter, are often dark, yet clays of warm-temperate and tropical regions with less than 3% organic content are some of the world's blackest soils.

To standardize color descriptions, soil scientists describe a soil's color by comparing it with a *Munsell Color Chart* (developed by artist and teacher Albert Munsell in 1913). These charts display 175 colors arranged by *hue* (H, the dominant spectral color, such as red), *value* (V, degree of darkness or lightness), and *chroma* (C, purity and saturation of the color, which increase with decreasing grayness).

Let's examine some essentials relative to soil color in four soil samples.

1. Using soil samples (observed/collected in the field or provided by your instructor, or soil photographs), note the predominant soil color and indicate the likely soil component responsible for the color. (Answers will vary, depending on the samples/photographs provided.) Be sure and note whether the sample is wet, moist, or dry.

   a) Soil sample A

   b) Soil sample B

   c) Soil sample C

   d) Soil sample D

Using These Same Four Samples

In the Munsell color system, every color has three qualities: *hue* (the dominant spectral color, such as red), *value* (degree of darkness or lightness), and *chroma* (purity and saturation of the color, which increase with decreasing grayness). The complete Munsell notation for a chromatic color is written symbolically, like this: H V/C. For example, for a strong red having a *hue* of 5R, a *value* of 6, and a *chroma* of 14, the complete Munsell notation is 5R 6/14 (5R is hue, 6 is value, and 14 is chroma). As another example, a pale brown is 10YR 6/3. A dark brown is noted as 10YR 2/2. More refined divisions of any of the attributes use decimals.

The light you use when you view a sample is important and can affect your assessment of the color notation. It is best to view the chart and the sample with the Sun over your shoulder, shining on the sample, with you facing away from the Sun. If you are in a lab or classroom under artificial light, try to have a light source as close to white light (all spectrum) as possible. Don't be too concerned at this stage; more practice will improve your assessment under varying light conditions. You will find low values and low chromas the most difficult to match against the color chips.

Note: When doing actual field work with a soil **pedon** (the complete **soil profile** and basic sampling unit in soil surveys), you will find different colors in each horizon, and maybe more than one color in a single horizon.

These details should be noted in your assessment.

2. Using a *Munsell Color Chart*, note the *color description* and the *Munsell notation* for each of the four samples you assessed in question 1.

   a) Soil sample A

   b) Soil sample B

   c) Soil sample C

   d) Soil sample D

SECTION 2

Soil Texture and Soil Structure

Soil texture refers to the mixture of sizes of its individual particles and the proportion of different sizes of soil separates (individual particles of soil). **Figure 28.1** illustrates particle size comparisons, and **Table 28.1** lists the standards for soil particle grades (sizes). Particles smaller than gravel are considered part of the soil, while larger particles such as gravel, pebbles, or cobbles are not. As you can see from the table, sand is further graded—ranging from very coarse to very fine.

If you have been to a beach, you have felt the texture of **sand**: It has a "gritty" feel. **Silt**, on the other hand, feels smooth—somewhat soft and silky, like flour used in baking bread. When wet, **clay** has a sticky feel and requires quite a bit of pressure to squeeze it, like the clay used in making pottery.

Soils nearly always consist of more than one particle size. By determining the relative amounts of sand, silt, and clay in a particular soil sample, it can be placed into one of 12 classes, as shown on the soil texture triangle in **Figure 28.2**. Each side presents percentages of a particle grade. See the line from each side of the triangle (following the direction indicated by the orientation of the numbers on each axis). You see that a soil consisting of 36% sand, 43% silt, and 21% clay is classified as **loam**, a term for soils consisting of mostly sand and silt with a relatively smaller amount of clay. This determination is a mechanical analysis or particle size analysis of the soil.

Soil structure refers to the arrangement of these soil separates. The smallest natural lump or cluster of particles is a **ped**. Soil structure is described as crumb or granular, platy, blocky, or prismatic or columnar.

Keep this distinction in mind when assessing soil: Soil separates of individual mineral particles are the *soil texture*, and peds, which are the arrangement of soil particles, comprise the *soil structure*.

TABLE 28.1 Soil texture grades		
Soil Particle Grade	Diameter (mm)	Diameter (in.)
Gravel	>2.0	>0.08
Very coarse sand	1.0–2.0	0.04–0.08
Coarse sand	0.5–1.0	0.02–0.04
Medium sand	0.25–0.5	0.01–0.02
Fine sand	0.10–0.25	0.004–0.01
Very fine sand	0.05–0.10	0.002–0.004
Silt	0.002–0.05	0.00008–0.002
Clay	<0.002	<0.00008

(U.S. Department of Agriculture, Natural Resources Conservation Service)

You can see in the table that soil grades from sand to silt to clay, from most-coarse to most-fine mineral particle sizes. Soils that represent the best particle size mix for plant growth are those that balance the three sizes. This is the loam described earlier in this section. Remember, pedology gives us a general understanding of soils and their classification, whereas edaphology reflects society's concern for food and fiber production and the management of soils to increase fertility and reduce soil losses.

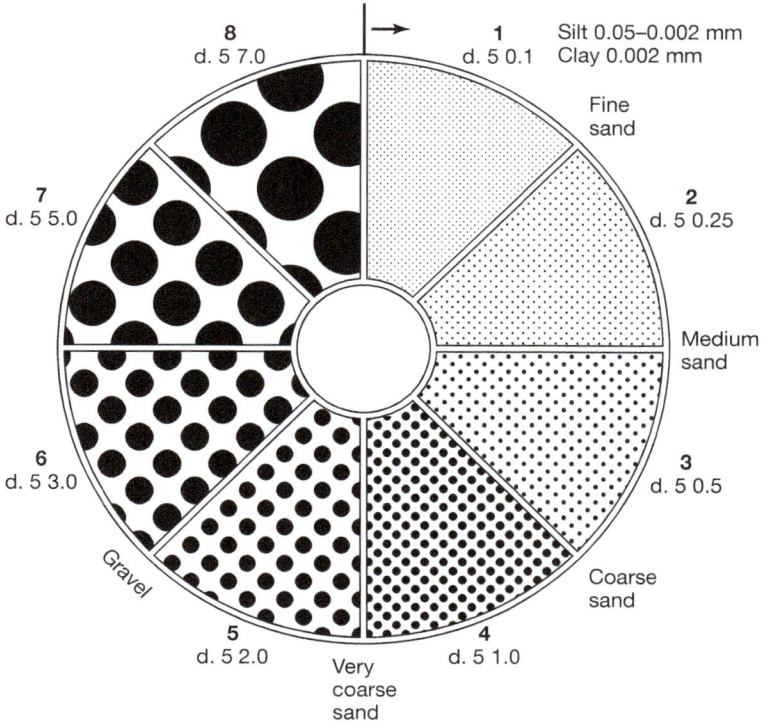

▲ **Figure 28.1** Sizes of soil particles larger than silt in approximate diameters (d) of millimeters (mm). (Adapted from *Soil Survey Manual*, USDA, Handbook No. 18, Natural Resources Conservation Service, October 1993, p. 137)

1. Use the soil texture triangle to name the following by its correct texture class (named on Figure 28.2):

   a) 17% sand, 28% silt, 55% clay: _____

   b) 31% sand, 55% silt, 14% clay: _____

2. Determine the percentage of each particle size for A and B examples plotted on the triangle (Figure 28.2):

   a) A: _____ % sand, _____ silt, _____ clay

   b) B: _____ % sand, _____ silt, _____ clay

Please read the appropriate section on soil horizons in *Geosystems*, *Elemental Geosystems*, or *Geosystems Core*. Soil textures vary through the horizons, so the soil texture triangle can be used for samples in each horizon. For instance, assume a silt loam in Ohio is sampled in the A, B, and C horizons. The mix of mineral particles occurring at each horizon gives you a complete sense of the soil. The textural analysis is as follows:

TABLE 28.2 Textural analysis of a silt loam soil			
Sample Points	% Sand	% Silt	% Clay
1 = A horizon	22	63	15
2 = B horizon	31	25	44
3 = C horizon	42	34	24

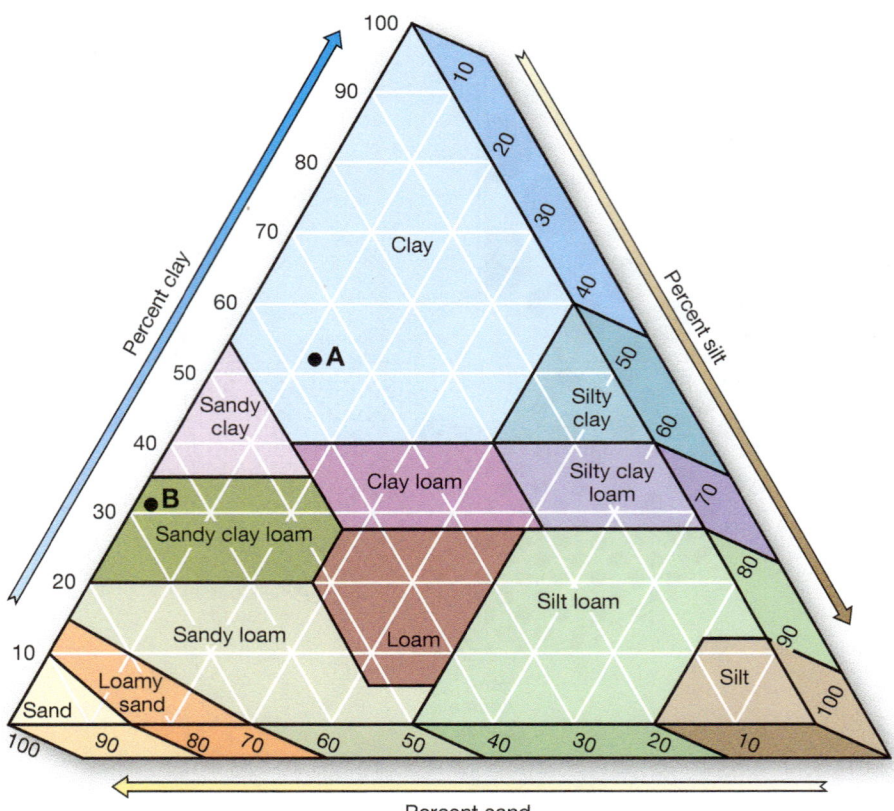

▲ **Figure 28.2** Soil texture triangle
A *Soil Survey Manual*, USDA, Handbook No. 18, Natural Resources Conservation Service, October 1993, p. 137.

3. Using the numbers **1, 2,** and **3** to designate each of the three soil horizons, place these numbers in the appropriate locations on the soil texture triangle in Figure 28.2. In the textural analysis of mineral particle sizes, the A horizon is mainly made up of

   the B horizon is mainly made up of

   and the C horizon is mainly made up of

   In a soil profile, is this what you would expect the particle size distribution to be? Explain.

4. In a less quantitative way than using sieves to physically sort soil particles, soil texture can be determined in the field by feeling the soil and estimating the percentages of sand, silt, and clay. Try this method using the following procedure with available soil samples, recording your observations and results through each of these steps. Your instructor may direct you to use the four original soil samples from previous sections of this exercise.

   Step 1: Fill your palm with dry soil, moistening it with enough water so that it sticks together sufficiently to be worked with your fingers. Add the water gradually: if it becomes too runny or if it sticks to your fingers, add more dry soil. You want a "plastic" mass that you can mold, somewhat like putty. (*Note:* If the soil is gritty and loose and won't form a ball but falls apart when rubbed between your fingers, it is a "sandy" soil.)

   Step 2: Knead the soil between your thumb and fingers, removing any pebbles that may be present and breaking up any peds (aggregates or clumps).

Step 3: The soil should form a ball when you squeeze it. Position the ball between your thumb and forefinger and gently push the ball with your thumb, squeezing it into a ribbon of uniform thickness and width. The ribbon will extend over your finger, eventually breaking from its own weight. Continue until several ribbons have formed and broken off. (If you have difficulty with this, roll the soil into a cylinder that is uniform in diameter and then squeeze the cylinder of soil into a ribbon.) The *clay content* will determine the length of the ribbon before it breaks:

- Ribbon length less than 1.5 in. (3.8 cm): clay content is <27%.
- Ribbon length 1.5 to 3 in. (3.8 to 7.6 cm): clay content is 27 to 40%.
- Ribbon length more than 3 in. (7.6 cm): clay content is >40%.

Step 4: Now put a small pinch of the soil sample into your palm and add enough water to excessively wet it. Rub it with your forefinger, *estimating* the sand content by the amount of *grittiness* you feel (silt has a smoothness, clay a stickiness in feel):

- Sand is the dominant texture you feel: sand content is >50%.
- Sand is noticeable, but not dominant: sand content is 20–50%.
- Predominant feel is smooth, indicating high silt content: sand content is <20%.

Step 5: Combine the estimates of sand and clay content and use **Table 28.3** to place your sample in one of the classes. (*Note:* If neither a gritty/sand nor smooth/silt feeling is predominant in Step 4, it will fall into the center column.)

Step 6: Record your observations in the spaces provided after Table 28.3. Your instructor will determine how many samples to use in your work. Space is provided here for three sample reports.

TABLE 28.3 Simplified soil texture table

		Estimated Percentage of Sand (predominant feeling)		
		>50 (gritty)	20–50 (neither)	<20 (smooth)
Estimated Percentage of Clay	>40	Sandy clay	Clay	Clay
				Silty clay
	27–40	Sandy clay loam	Clay loam	Silty clay loam
	<27	Sandy loam	Loam	Silt loam
		Loamy sand		
		Sand		

Post your results from **Step 4** and **Step 5** and Table 28.3 here:

a) Soil sample A

b) Soil sample B

c) Soil sample C

Note that in soil science, the term *consistence* is used to describe the consistency of a soil or cohesion of its particles. Consistence is a product of texture (particle size) and structure (ped shape). You are working with soil consistence in the exercise above. Consistence reflects a soil's resistance to breaking and manipulation under varying moisture conditions:

- A wet soil is sticky between the thumb and forefinger, ranging from a little adherence to either finger, to sticking to both fingers, to stretching when the fingers are moved apart. *Plasticity*, the quality of being moldable, is roughly measured by rolling a piece of soil between your fingers and thumb to see whether it rolls into a thin strand.
- A moist soil is filled to about half of field capacity (the usable water capacity of soil), and its consistence grades from loose (noncoherent), to friable (easily pulverized), to firm (not crushable between thumb and forefinger).
- A dry soil is typically brittle and rigid, with consistence ranging from loose, to soft, to hard, to extremely hard.

SECTION 3

Soil Texture and Porosity

Soil texture affects the movement of water through the soil and, therefore, the amount of water available to plants. Sandy soils have large air spaces between the particles, making them very porous, allowing the water to percolate through the soil quickly. Thus, little water is retained, causing the soil to dry out quickly. You may have observed this at a beach as water quickly drained through the sand, leaving the surface dry and loose.

The small particles in clay soils fit closely together, leaving few air spaces between them so that water percolates very slowly, collecting on the soil surface or running off if there is any slope. Once the water is absorbed into the clay soil, it is held so tightly that little is available to the plants.

Therefore, the texture and the structure of the soil dictate available pore spaces, or porosity. The property of the soil that determines the rate of soil-moisture recharge is its **permeability**. Permeability depends on particle sizes and the shape and packing of soil grains.

Loam soils have a loose, porous texture due to their mixture of particle sizes. As a result, loams retain water well without becoming waterlogged, leaving the water readily available to plants. This is one reason why farmers consider loamy soils ideal. **Table 28.4** presents water absorption rates for various soil textures.

TABLE 28.4 Water absorption rates in soil	
Soil Texture	**Absorption Rate per Hour**
Sandy	>5 cm (+2 in.)
Loam	0.6–5 cm (0.25–2 in.)
Clay	<0.6 cm (<0.25 in.)

1. You can perform a simple demonstration to observe the effect of soil texture and structure on porosity. Your instructor may have you use the same soil samples from earlier sections in this exercise.

   Step 1: Obtain samples of several soils of differing textures and put each in a glass jar or beaker. Fill the jars to the same level, leaving several inches at the top, taking care not to compact the soil in the container.

   Step 2: Measure 5 cm (2 in.) of water in another container of the same size. Measure the same amount for each sample. Pour the water over the soil in each container and observe how quickly the water percolates through the soil to the bottom of the jar. Roughly time the percolation rates. (*Note:* The main focus of this demonstration is observation, not exact measurement and quantification of the percolation rates.)

Step 3: Use the spaces below to make a notation of your observations. Room for three soil sample observations is given. Assess the texture of each sample, based on these observations.

a) Soil sample A

b) Soil sample B

c) Soil sample C

2. Optional: A similar demonstration can be done "in the field." Selecting a site where the soil is not compacted, trim any vegetation completely to the ground and carefully remove any loose organic materials. Pound a soup can (both ends removed) into the soil several inches. If necessary, place a board over the top of the can to prevent the can from crumpling while you do this. Pour water into the can, filling it to the top. After a period of time, observe how much water has been absorbed. You can do this demonstration at several sites and compare the results. (Again, while you could take careful measurements, the purpose of this investigation is primarily observation and relative absorption rates.)

a) Site A

b) Site B

c) Site C

SECTION 4

Soil Profiles of Soil Horizons

Just as a book cannot be judged by its cover, so soils cannot be evaluated at the surface only. Instead, a soil profile should be studied from the surface to the deepest extent of plant roots, or to where regolith or bedrock is encountered. Such a profile, called a pedon, is a hexagonal column measuring 1–10 m² in top surface area. At the sides of the pedon, the various layers of the soil profile are visible in cross section and are labeled with letters. *A pedon is the basic sampling unit used in soil surveys.*

Many pedons together in one area make up a **polypedon**, which has distinctive characteristics differentiating it from surrounding polypedons. A polypedon is comprised of an identifiable series of soils in an area. It can have a minimum dimension of about 1 m² and no specified maximum size. *The polypedon is the basic mapping unit used in preparing local soil maps.*

Each distinct layer exposed in a pedon is a **soil horizon**. Each layer or horizon is distinct from the one directly above or below, with visible boundaries distinguished by differences in soil properties, some of which are color, texture, porosity, moisture, and the presence or absence of certain moisture, and the presence or absence of certain minerals or other materials. The group of horizons from the O horizon down through the B horizon together are the **solum**, considered the true soil profile.

▲ **Figure 28.3** Ideal soil profile

Figure 28.3 illustrates the horizons in a model or ideal profile. From the surface down the soil horizons are as follows:

- **O horizon**—Named for the organic materials from plant and animal litter. This layer is often subdivided into two layers: the topmost is O_1, consisting of leaves and other largely undecomposed organic material; just beneath is O_2, which is the decomposed organic debris called **humus**.
- **A horizon**—The uppermost mineral horizon. This dark-colored layer is rich in fine clay-sized particles and organic material derived from the humus above as well as plant roots.
- **E horizon**—Layer named for the process of *eluviation*, in which clays, aluminum and iron oxides, and organic matter are leached (washed out) to lower layers. This horizon is lighter in color and comprised largely of sand and coarse silt particles.
- **B horizon**—Characterized by the accumulation of materials leached from above. This deposition process is called *illuviation*. Presence of oxides may give this layer reddish or yellowish hues, and the accumulated fine clay and organic particles make this layer quite dense.
- **C horizon**—Not considered part of true soil. Consisting of **regolith**, or weathered parent material, this layer lacks biological activity.
- **R horizon**—The bottom layer of the profile, and like the C horizon above, not considered to be part of true soil. This horizon is either unconsolidated (loose) material or solid bedrock.

The Illustrated Guide to Soil Taxonomy, available for download at https://www.nrcs.usda.gov/wps/portal/nrcs/detail/soils/survey/class/taxonomy/?cid=nrcs142p2_053580 has many images of soil horizons.

1. Using photographs of three soil profiles (either in the soils chapter of a textbook or provided by your instructor) and working in groups with your lab partners, identify the various soil horizons in each profile. Observe horizon characteristics such as thickness, color, texture, and structure, and note variations from one profile to another. Speculate on the factors that combined to produce the appearance of each.

   a) Soil profile A

   b) Soil profile B

   c) Soil profile C

2. You may be able to locate one or more sites on your campus or nearby where you can see soil profiles in their natural setting. As you did in the first activity, observe the profile's characteristics; if more than one profile is available, note variations between them and see if you can determine the major factors influencing the formation of each.

   a) Profile site A

   b) Profile site B

   c) Profile site C

SECTION 5

Soil Acidity and Alkalinity

Soil pH, the **acidity** or **alkalinity** of the soil as expressed on the pH scale, strongly affects soil fertility (**Figure 28.4**). Nutrient availability is low in soils that are either very acidic or very alkaline. A soil rich in hydrogen ions (H+) is an acid soil. On the other hand, a soil high in base cations (calcium, magnesium, potassium, sodium) is a basic, or alkaline, soil.

Pure water is nearly neutral, with a pH of 7.0. Readings below 7.0 represent increasing acidity. Readings above 7.0 indicate increasing alkalinity. Acidity usually is regarded as strong at 5.0 or lower, whereas 10.0 or above is considered strongly alkaline.

Several factors influence soil acidity. The chemistry of soil parent materials, as well as crop fertilization and harvesting, can increase soil acidity. However, the major contributor to soil acidity in this modern era is acid precipitation (rain, snow, fog, or dry deposition). Acid rain actually has been measured below pH 2.0—an incredibly low value for natural precipitation, as acid as lemon juice. Increased acidity in the soil solution accelerates the chemical weathering and depletion rates of some mineral nutrients, and it can also decrease the availability of other nutrients. Because most crops

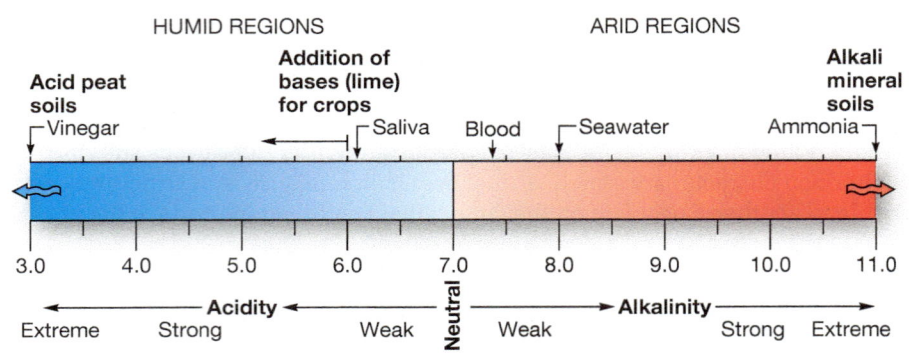

▲ **Figure 28.4** pH scale

are sensitive to specific pH levels, acid soils below pH 6.0 require treatment to raise the pH. This soil treatment is accomplished by the addition of bases in the form of minerals that are rich in base cations, usually lime (calcium carbonate, $CaCO_3$).

Soil pH is regularly tested at many locations using an *electrometric method* (a probe into a soil-water mixture) or using a *dye method* (organic compounds in drops or test paper that react and indicate pH through color changes). More accurate work can be done in a laboratory. Care must be taken in the field when reading the pH of a soil sample because values may vary over a short distance, perhaps in response to previous fertilizer applications, or reflect topographic effects. Despite these variations, soil pH is related to so many biological and chemical factors that influence both soil processes and plant response that it is an important parameter to know.

1. Using pH test equipment provided by your instructor, evaluate two soil samples, perhaps one from your home and one from campus, or from two different areas on your campus. Ideally, you would use samples from sites with different vegetation. Record their pH readings on Figure 28.4 above. How far apart on the pH scale are the two readings?

2. Are either of the samples strongly acidic? What remedial actions could be taken to make them more pH neutral, and under what circumstances might you want to do this?

3. Review Focus Study 3.1 in *Geosystems* or Focus Study 2.1 in *Elemental Geosystems*. How has acid deposition affected soils, forests, and lakes in the United States and Europe? Do you think that the remedial actions you outlined in item 2 are practical for large-scale problems affecting entire regions? What do you think would be the best solution at these larger spatial scales? (*Hint:* Consider what the sources of acidifying materials are.)

SECTION 6

In the **soil classification** system developed by the Natural Resources Conservation Service, there is a hierarchy of soil categories. **Table 28.5** presents the categories of the U.S. Soil Taxonomy.

TABLE 28.5 U.S. soil taxonomy	
Soil Category	Number of Soils Included
Orders	12
Suborders	47
Great groups	230
Subgroups	1200
Families	6000
Series	15,000

1. Use Figure 28.5, map of the worldwide distribution of the Soil Taxonomy orders, find your present approximate location. Although this is only a general assessment because of the scale of the world map, use the color map legend and identify the one or two soil orders and their main characteristics that seem to represent your region. Record your findings here:

Web page
Soil
Surveys

https://goo.gl/a8ChLn

Optional—Internet access required

2. You can see from Table 28.5 that there are some 15,000 soil series identified in the soil taxonomy for the United States and Canada. The USDA-Natural Resources Conservation Service has a website about its soil surveys: http://websoilsurvey.sc.egov.usda.gov/App/HomePage.htm.

To find out about the horizons, color, texture, structure, consistence, mineral, and chemical composition of your soil—or any other soil—go to this website.

The website allows you to set an area of interest and explore it in amazing detail. Zoom in to your school and use the Area of Interest tool to select your campus. The Soil Map tab displays all the individual soils found within the Area of Interest you set as an overlay on top of the satellite image of your campus, as well as a clickable list. The list of map units provides a wealth of data, including elevation, mean annual precipitation and temperature, frost-free period, farmland classification, landforms associated with that map unit, slope, drainage class, depth to water table, and water capacity. The Soil Data Explorer tab allows you to research the suitablity of soils for various uses, such as building site development, construction materials, vegetative productivity, and water management.

◀ **Figure 28.5** Worldwide Distribution of the Soil Taxonomy orders

Zoom in to your campus and use the tools to set the Area of Interest to your campus.

Answer the following questions for your campus: how many map units are on your campus? Which covers the largest area? How many acres and what percent of your campus does it cover? How well drained is it? What is its farmland classification?

Note: When doing actual field work with a soil pedon (the complete soil profile and the basic sampling unit in soil surveys), you will find different colors in each horizon and perhaps a couple colors in one horizon. If you identify more than one color in the soil samples provided for this exercise, be sure to note it in your assessment.

Mastering Geography

Looking for additional lab prep and review materials? Visit the Study Area in Mastering Geography to enhance your geographic literacy and spatial reasoning skills by accessing MapMaster™ interactive maps, videos, *Mobile Field Trips, Project Condor* Quadcopter Videos, *In the News* articles, flashcards, web links, self-study quizzes, and other resources. Contact your professor to find out how you can access Mastering Geography.

NOTES

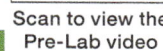

Video
Exercise 29
Pre-Lab Video

http://goo.gl/DVClyQ

Scan to view the
Pre-Lab video

LAB EXERCISE

29 Biomes: Analyzing Global Terrestrial Ecosystems

The Earth systems science approach embodied in physical geography synthesizes content from across the disciplines to create a holistic perspective. Exciting progress toward an integrated understanding of Earth's physical and biological systems is under way. Physical geography plays an important role in mapping and analyzing Earth's **terrestrial ecosystems**.

The diversity of organisms is a response to the interaction of the atmosphere, hydrosphere, and lithosphere, producing diverse conditions within which the biosphere exists. A first-ever international attempt to protect biodiversity (species richness) is now ratified through the auspices of the United Nations.

The biosphere includes myriad ecosystems from simple to complex, each operating within general spatial boundaries. An **ecosystem** is a self-regulating association of living plants and animals and their nonliving physical environment. In an ecosystem, a change in one component causes changes in others, as systems adjust to new operating conditions. Interacting populations of plants and animals in an area form a **community**. Each plant and animal occupies an area in which it is biologically suited to live—its habitat—and within that **habitat** it performs a basic operational function—its **niche**.

Earth itself is the largest ecosystem within the natural boundary of the atmosphere. Natural ecosystems are open systems for both energy and matter, with almost all ecosystem boundaries functioning as transition zones rather than as sharp demarcations.

Plants are the most visible part of the biotic landscape, a key aspect of Earth's terrestrial ecosystems. In their growth, form, and distribution, plants reflect Earth's physical systems: its energy patterns; atmospheric composition; temperature and winds; air masses; water quantity, quality, and seasonal timing; soils; regional climates; geomorphic processes; and ecosystem dynamics. The net photosynthesis for an entire plant community (photosynthesis minus respiration) is its *net primary productivity*. This is the amount of stored chemical energy (biomass) that the community generates for the ecosystem. **Biomass** is the net dry weight of organic material and varies among ecosystems.

A large, stable terrestrial ecosystem is known as a **biome**. Specific plant and animal communities and their interrelationship with the physical environment characterize a biome. Each biome is usually named for its *dominant vegetation*. We can generalize Earth's wide-ranging plant species into six broad biomes: *forest, savanna, grassland, shrubland, desert*, and *tundra*. Because plant distributions are responsive to environmental conditions and reflect variations in climatic and other abiotic factors, the world climate map and the global terrestrial biome map are presented together on the back cover of this lab manual.

Earth's diversity is expressed in 300,000 plant species. Despite this complexity of diverse plant and animal communities and their interrelationships, we can generalize Earth's ecosystems into 10 global terrestrial biome regions on the global terrestrial ecosystem map (on the back cover of this lab manual) and in the comprehensive **Table 29.1**. The table synthesizes many aspects of physical geography, integrating them under the 10 biomes. These biomes are included in the glossary and are presented on the biome map and integrative table. Lab Exercise 29 features mobile-friendly QR links to Google Earth virtual tours that fly you to the topographic map location, as well as optional online Google Earth activities with downloadable media and worksheets at mygeoscienceplace.com and the Mastering Geography Study Area.

Key Terms and Concepts

arctic tundra
biodiversity
biogeography
biomass
biome
cold desert and semidesert
community
ecosystem
equatorial and tropical rain forest

formation class
habitat
island biogeography
life zone
Mediterranean shrubland
midlatitude broadleaf and
   mixed forest
midlatitude grasslands
montane forest

needleleaf forest
niche
temperate rain forest
terrestrial ecosystem
tropical savanna
tropical seasonal forest and
   scrub
warm desert and semidesert

KEY**LEARNING**concepts

After completion of this lab, you should be able to:

- **Identify** and *differentiate* various terrestrial ecosystem formation classes as they relate to patterns of precipitation and temperature.

- **Compare** Köppen climate classifications from Lab Exercise 17 with the pattern of terrestrial ecosystems (see Appendix B, *Geosystems*, 10/e, or Appendix C in *Elemental Geosystems*, 9/e).

- **Relate** vertical life zones to the latitudinal distribution of biomes.

- **Utilize** a world biome map and an integrative table of major terrestrial biomes and their characteristics to *analyze* environmental conditions.

Materials/Sources Needed

pencil
Google Earth

world atlas or physical geography text

Lab Exercise and Activities

SECTION 1

Climate Controls of Ecosystem Structure and Form

Figure 29.1 illustrates the general relationship among temperature, precipitation, and vegetation. **Formation classes** are units that refer to the structure and appearance of dominant plants in a terrestrial ecosystem, for example, forest, shrubland, or grassland. Each formation includes numerous plant communities, and each community includes innumerable plant habitats. The illustration relates temperature and precipitation over hot, temperate, cool, and polar and alpine environments. As you examine the illustration, mentally review what you learned about the characteristics and distribution of climates in Lab Exercise 17.

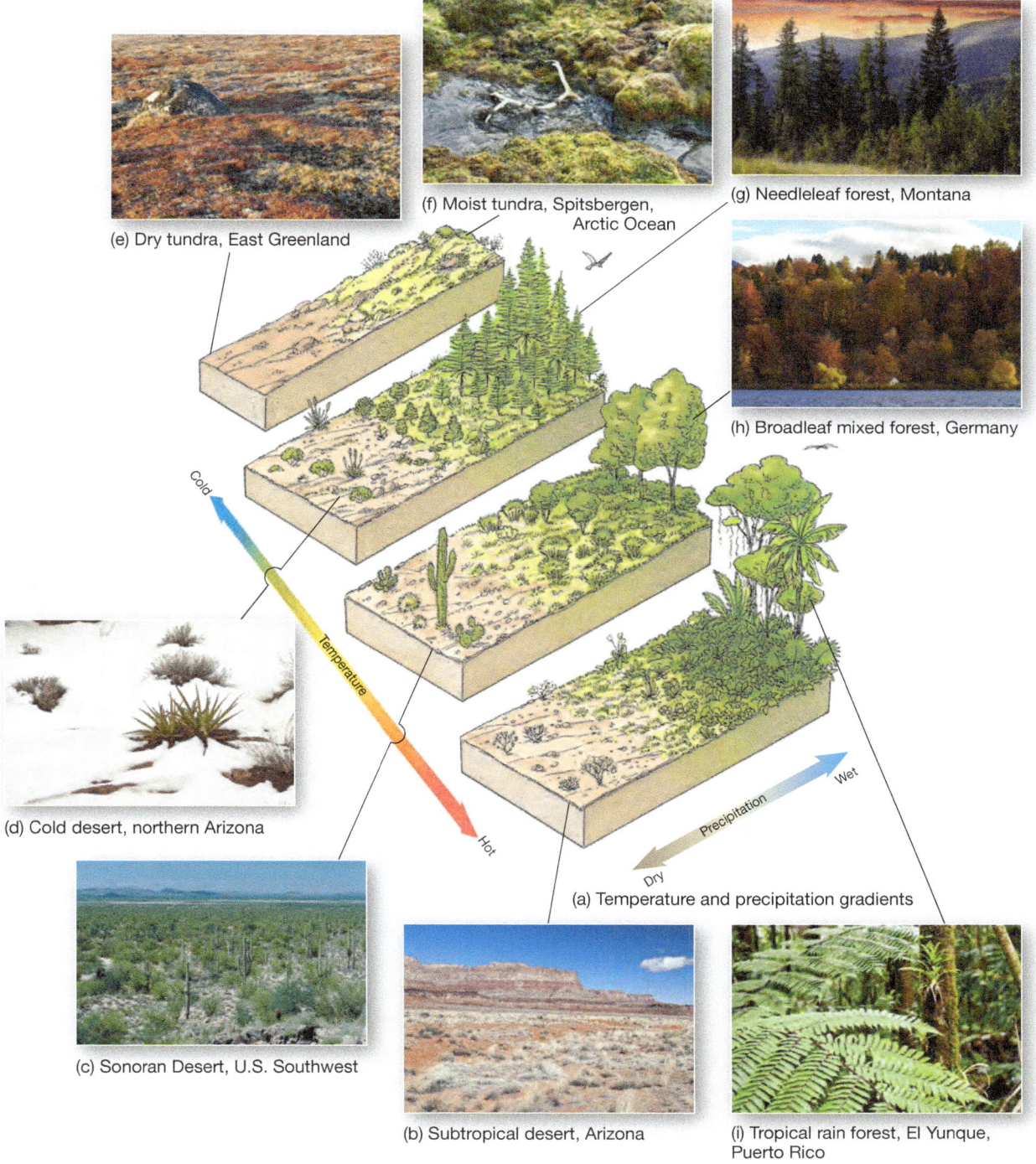

(e) Dry tundra, East Greenland

(f) Moist tundra, Spitsbergen, Arctic Ocean

(g) Needleleaf forest, Montana

(h) Broadleaf mixed forest, Germany

Cold

Temperature

Hot

(d) Cold desert, northern Arizona

Precipitation

Wet

Dry

(a) Temperature and precipitation gradients

(c) Sonoran Desert, U.S. Southwest

(b) Subtropical desert, Arizona

(i) Tropical rain forest, El Yunque, Puerto Rico

▲ **Figure 29.1** Abiotic (nonliving) climate controls of ecosystem types: the generalized relationship among rainfall, temperature, and vegetation. [(b), (d)–(f), and (i) Bobbé Christopherson. (c) Robert Christopherson. (g) Snehit/Shutterstock. (h) blickwinkel/Hartl/Alamy.]

Questions using Figure 29.1:

1. Describe in general terms the characteristic vegetation type and related temperature and moisture relationship that fit the area of your present town or school.

   Now, describe the same for the region where you were born (if substantially different).

2. Using Figure 29.1, and referring to the world biome map (on the back cover of this lab manual), write these city names in the appropriate environmental location—locate and label each one on the figure. Discuss these with other lab members to make your determinations.

   a) Your present region
   b) Where you were born (if a different region from your present location)
   c) New York City
   d) Key West, Florida
   e) Montreal, Québec
   f) Dawson, Yukon
   g) Yuma, Arizona
   h) Omaha, Nebraska
   i) Elko, Nevada
   j) Everglades, Florida
   k) Coastal Oregon

   Recall that Köppen's climatic classification was based on two climate control regimes:

1. **Temperature**—reflecting either

   • Latitude—as in   • Altitude—as in *highland*
   tropical   *or*
   subtropical
   midlatitude
   subarctic/subpolar
   polar
   icecap

2. **Precipitation**—as in
   rainy
   wet and dry
   monsoon
   arid
   semi-arid
   humid
   dry

   In addition, global pressure and wind belts influence precipitation regimes:

   • **Rainy/humid conditions**—result from one of the following:
     • instability and convectional uplift in warm *equatorial* and *subtropical* regions
     • frontal uplift in the *midlatitudes*
     • intensification of rainfall on the *windward side* of mountains

   • **Dry conditions**—caused by one of the following:
     • subsiding air of the *subtropical high pressure belts*
     • isolation within *midlatitude* continental interiors
     • cold stable air of the *high latitudes*
     • drying of air masses descending the *leeward side* of mountain ranges

These are broad generalizations but can help you to see the causes of resulting climatic patterns. You will refer to these relationships in Section 3.

The descriptive names in Köppen's climate classification system (see Appendix B in *Geosystems*, or Appendix C in *Elemental Geosystems*) also often included the dominant vegetation types that those regimes supported: *Tropical Rainy or Rainforest, Tropical Wet and Dry or Savanna, Tropical/Midlatitude Arid or Desert, Tropical/Midlatitude Semi-arid or Steppe, Subtropical Dry or Mediterranean, Subarctic/Subpolar or Taiga, Polar or Tundra.*

SECTION 2

Earth's Major Terrestrial Biomes

The global distribution of Earth's major terrestrial biomes is portrayed on the map on the back cover of this manual. Table 29.1 describes each biome on the map and summarizes other pertinent environmental information—a compilation from many aspects of physical geography, for Earth's biomes are a synthesis of the environment and biosphere.

Questions and completion items about this sample of Earth's terrestrial biomes.

1. From Table 29.1, the biome map, and the climate map, determine the terrestrial biome that best characterizes each of the following descriptions and write its name *on the first line* provided. *On the second line,* relate it to temperature and precipitation regimes and/or global pressure and wind belts outlined at the end of Section 1. The first one is completed for you as an example.

    a) PRECIP less than 1/2 POTET: *[Warm desert and semidesert]*

                        [midlatitude continental interior and on leeward side of mountain range]

    b) Southern and eastern U.S. evergreen pines:

    c) Mollisols and aridisols:

    d) Characteristic of central Australia:

    e) Selva:

    f) Characteristic of the majority of central Canada:

    g) Transitional between rain forest and tropical steppes:

    h) Tallest trees on Earth:

    i) Sedges, mosses, and lichens:

    j) Characteristic of Zambia (south-central Africa):

    k) Four biome types that occur in Chile:

    l) Precipitation of 150–500 cm/year, outside the tropics:

    m) Characteristic of central Greenland:

TABLE 29.1 Major terrestrial biomes and their characteristics

Biome and Ecosystems (map symbol)	Vegetation Characteristics	Soil Orders (Soil Taxonomy)	Köppen Climate Designation	Annual Precipitation Range	Temperature Patterns	Water Balance
Equatorial and Tropical Rain Forest (ETR) Evergreen broadleaf forest Selva	Leaf canopy thick and continuous; broadleaf evergreen trees, vines (lianas), epiphytes, tree ferns, palms	Oxisols Ultisols (on well-drained uplands)	Af Am (limited dry season)	180–400 cm (>6 cm/mo)	Always warm (avg 25°C)	Surpluses all year
Tropical Seasonal Forest and Scrub (TrSF) Tropical monsoon forest Tropical deciduous forest Scrub woodland and thorn forest	Transitional between rain forest and grasslands; broad-leaf, some deciduous trees; open parkland to dense undergrowth; acacias and other thorn trees in open growth	Oxisols Ultisols Vertisols (in India) Some alfisols	Am Aw Borders BS	130–200 cm (>40 rainy days during 4 driest months)	Variable, always warm (>18°C)	Seasonal surpluses and deficits
Tropical Savanna (TrS) Tropical grassland Thorn tree scrub Thorn woodland	Transitional between seasonal forests, rain forests, and semiarid tropical steppes and desert; trees with flattened crowns, clumped grasses, and bush thickets; fire association	Alfisols (dry: Ultalfs) Ultisols Oxisols	Aw BS	90–150 cm, seasonal	No cold-weather limitations	Tends toward deficits, therefore fire- and drought-susceptible
Midlatitude Broadleaf and Mixed Forest (MBMF) Temperate broadleaf Midlatitude deciduous Temperate needleleaf	Mixed broadleaf and needleleaf trees; deciduous broadleaf, losing leaves in winter; southern and eastern evergreen pines demonstrate fire association	Ultisols Some alfisols	Cfa Cwa Dfa	75–150 cm	Temperate, with cold season	Seasonal pattern with summer maximum PRECIP and POTET (PET); no irrigation needed
Needleleaf Forest and Montane Forest (NF/MF) Taiga Boreal forest Other montane forests and highlands	Needleleaf conifers, mostly evergreen pine, spruce, fir; Russian larch, a deciduous needleleaf	Spodosols Histosols Inceptisols Alfisols (boralfs: cold)	Subarctic Dfb Dfc Dfd	30–100 cm	Short summer, cold winter	Low POTET (PET), moderate PRECIP, moist soils, some water-logged and frozen in winter; no deficits
Temperate Rain Forest (TeR) West Coast forest Coast redwoods (U.S.)	Narrow margin of lush evergreen and deciduous trees on windward slopes; redwoods, tallest trees on Earth	Spodosols Inceptisols (mountainous environs)	Cfb Cfc	150–500 cm	Mild summer and mild winter for latitude	Large surpluses and runoff
Mediterranean Shrubland (MSh) Sclerophyllous shrubs Australian eucalyptus forest	Short shrubs, drought adapted, tending to grassy woodlands; chaparral	Alfisols (Xeralfs) Mollisols	Csa Csb	25–65 cm	Hot, dry summers, cool winters	Summer deficits, winter surpluses
Midlatitude Grasslands (MGr) Temperate grassland Sclerophyllous shrub	Tallgrass prairies and short-grass steppes, highly modified by human activity; major areas of commercial grain farming; plains, pampas, and veld	Mollisols Aridisols	Cfa Dfa	25–75 cm	Temperate continental regimes	Soil moisture utilization and recharge balanced; irrigation and dry farming in drier areas
Warm Desert and Semidesert (DBW) Subtropical desert and scrubland	Bare ground graduating into xerophytic plants including succulents, cacti, and dry shrubs	Aridisols Entisols (sand dunes)	BWh BWk	<2 cm	Average annual temperature around 18°C, highest temperatures on Earth	Chronic deficits, irregular precipitation events, PRECIP <½ POTET (PET)
Cold Desert and Semidesert (DBC) Midlatitude desert, scrubland, and steppe	Cold desert vegetation includes short grass and dry shrubs	Aridisols Entisols	BSh BSk	2–25 cm	Average annual temperature around 18°C	PRECIP >½ POTET (PET)
Arctic and Alpine Tundra (AAT)	Treeless; dwarf shrubs, stunted sedges, mosses, lichens, and short grasses; alpine, grass meadows	Gelisols Histosols Entisols (permafrost)	ET Dwd	15–80 cm	Warmest months <10°C, only 2 or 3 months above freezing	Not applicable most of the year, poor drainage in summer
Ice			EF			

n) Characteristic of Iran (northeast of the Persian Gulf):

o) Major area of commercial grain farming:

p) Seasonal precipitation of 90 to 150 cm/year:

q) Cfa and Dfa climate types:

r) Spodosols and permafrost, short summers:

s) Southern Spain, Italy, and Greece, central California:

t) Characteristic of Ireland and Wales:

u) Just west of the 98th meridian in the United States:

v) Just east of the 98th meridian in the United States:

w) Bare ground and xerophytic plants:

x) East coast of Madagascar:

y) West coast of Madagascar:

z) Characteristic of northern Mexico:

Finally, from the **GEOGRAPHY I. D.** that you completed in the Preface of this lab manual, complete the following for your hometown or college campus location:

Place name: Köppen classification:

2. What are the main climatic influences for this station (air pressure, air mass sources, degree of continentality, temperature of ocean currents)?

Terrestrial biome characteristics:

SECTION 3

Life Zones—Conditions Changing with Elevation

Alexander von Humboldt (1769–1859)—an explorer, geographer, and scientist—deduced that plants and animals occur in related groupings wherever similar conditions occur in the abiotic environment. After several years of study in the Andes Mountains of Peru, he described a distinct relationship between elevation and plant communities, his **life zone** concept. As he climbed the mountains, he noticed that the experience was similar to that of traveling away from the equator toward higher latitudes.

This zonation of plants with altitude is noticeable on any trip from lower valleys to higher elevations. Each life zone possesses its own temperature, precipitation, and insolation relationships and, therefore, its own biotic communities. The key is that temperatures decrease rapidly with increasing elevation at an average of 6.4 C° per kilometer (3.5 F° per 1000 ft), a rate known as the *normal lapse rate.*

The Grand Canyon in Arizona provides a good example. The inner gorge at the bottom of the canyon (600 m, or 2000 ft, in elevation) exhibits life forms characteristic of the lower Sonoran Desert of northern Mexico. However, the north rim of the canyon (2100 m, or 7000 ft, in elevation) is dominated by ecosystems similar to those of southern Canadian forests. On the summits of the nearby San Francisco Mountains (3600 m, or 12,000 ft, in elevation), the vegetation is similar to the arctic tundra of northern Canada.

1. Using a physical geography text, such as *Geosystems*, add appropriate labels to **Figure 29.2** in the spaces provided at the bottom of Figure 29.2: tropical rain forest, temperate deciduous forest, needleleaf forest, tundra, and ice and snow. Next, assuming the normal lapse rate is in effect, beginning at sea level and the rain forest label, assume the temperature to be at 27°C (80°F). Assume the labels are spaced at 600m (2000 ft) intervals and calculate the temperature on this particular day at each label elevation. Calculate five temperatures decreasing with altitude (600, 1200, 1800, 2400, and 3000 m) and record your answers on the lines along the right side of the mountain in Figure 29.2.

2. Following discussion among lab members, compare climate controls (temperature and precipitation) and the effects of altitude—Figures 29.1 and 29.2. Briefly describe the importance of each control on ecosystem character and plants and animals.

   a) Temperature:

   b) Precipitation:

   c) Elevation:

Vertical zonation
Increasing elevation from sea level

Latitudinal zonation
Increasing latitude from the equator

▲ **Figure 29.2** Progression of generalized plant community life zones with increasing elevation or latitude.

3. Relative to the life zone concept in Bolivia, South America, speculate on these items:

   a) Given these concepts of elevation and temperature, why are there glaciers along the crest of the Andes Mountains so near the equator?

   b) Why are Bolivians able to grow wheat, potatoes, and barley at 4250 m (14,000 ft) at the same latitude of rain forests at low elevation where these same crops will not grow?

SECTION 4

Island Biogeography

Biogeography is the study of the temporal and spatial distribution of plants and animals. Biogeographers study aspects of a population of individuals of a species, as well as general patterns of biological communities. Biogeography can be used across a wide range of spatial and temporal scales, from a group of individuals in one particular study area to climate and geological factors operating at a global spatial scale and geologic time scale. It draws upon the fields of ecology, evolutionary biology, geology, physical geography, and GIS (Geographic Information Systems). Biogeographers have worked to identify factors that cause species and biological communities to vary with latitude, elevation, insolation, and habitat size. Patterns of species distributions are often understood through speciation and extinction, plate tectonics, climate change (including glaciation and changes in sea level that expose and submerge land bridges), habitat size and isolation, and insolation, which powers photosynthesis.

David Quammen described biogeography as practiced by scientists in his book *The Song of the Dodo*: "Biogeography does more than ask *Which species?* and *Where?* It also asks *Why?* and, what is sometimes even more crucial, *Why not?*"

An important part of biogeography is **island biogeography**, the study of the factors that determine the number of species in a specific community. An island is not just a location surrounded by water but can be any location surrounded by a significantly different habitat or barrier to dispersal. Island biogeography is a very powerful concept because it applies to many situations. Loss of **biodiversity**, or species richness, is an important concern in ecology since the present rate of the loss of species is equaled only by mass extinction events that in the past have been caused by global catastrophes, such as the Cretaceous–Tertiary extinction event that saw the end of 75% of all the species on Earth at the time.

Scientists have been aware of the relationship between habitat size and the number of different species, called the species–area relationship. The species–area relationship is not linear, but logarithmic, because for the number of species to double, the habitat area needs to increase by 10 times.

In 1962 Robert MacArthur and Edward O. Wilson published *The Theory of Island Biogeography*, which refined the species–area relationship. They considered the size of an island, and also the distance of the island from the mainland, which is the source of new individuals and species. Larger islands have several advantages over smaller islands, in terms of biodiversity. Larger islands are bigger targets, so more individuals will potentially find them. Larger islands also may have more different habitats. MacArthur and Wilson hypothesized that an equilibrium biodiversity value could be calculated, based upon the rate of new species arriving and the rate of extinction for species on the island. **Figure 29.3** shows four different predicted equilbrium levels based on the size of the island and its distance from the mainland. One application of island biogeography is studying the effects of habitat fragmentation on species loss in order to design natural reserves to preserve biodiversity.

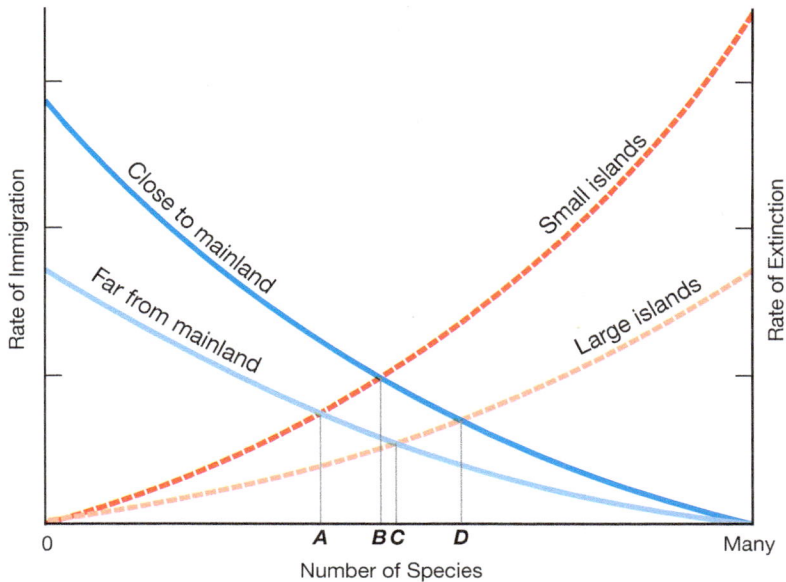

▲ **Figure 29.3** Equilibrium theory of island biogeography [Based on MacArthur, R. H., and E. O. Wilson. *The Theory of Island Biogeography*. 1967. Princeton University Press.]

1. Figure 29.3 shows four examples of species equilibrium numbers. In this hypothetical example, which combination of factors produces the highest number of species? The lowest number of species? In this example, is island size or distance from the mainland more important in producing a higher species equilibrium number?

2. What are some of the characteristics of animals that you would expect to find on an island? What are some characteristics of animals that you would not expect to find?

3. Some islands, such as the Hawaiian Islands, are far from the nearest land and were initially bare rock. Other islands, such as the Channel Islands, off the southern California coast, were part of the mainland and were isolated by rising sea levels. How might the organisms on the Hawaiian Islands differ from organisms found on the Channel Islands?

MacArthur and Wilson's theory of equilibrium species numbers took into account new species and organisms colonizing the islands from a mainland. There are, however, locations where the number of new immigrants may be zero. High mountain habitats in the western United States are one example. As climate became warmer and drier during the Pleistocene, habitats shifted up in elevation in response, creating islands of montane habitat surrounded by large expanses of desert. The desert is an impassible barrier for nonflying animals. In this example, there is no mainland source of new individuals and species. Lomolino, Brown, and Davis examined biodiversity in montane forests in the western United States. Their results are shown in **Table 29.2**.

	Forest Island	Species	Area (km²)
	TABLE 29.2 Species numbers and habitat size for sky islands (from Lomolino, M. V., J. H. Brown, and R. Davis. 1989. Island biogeography of montane forest mammals in the American Southwest. *Ecology* 70:180–194)		
1	Mogollon	16	11134.33
2	Lasals	13	435.23
3	Chuskas	11	1507.9
4	Abajos	10	958.57
5	Sandias	10	43.85
6	Sacramentos	10	1349.8
7	Mt. Taylor	8	465.3
8	Manzanos	8	141.13
9	Blacks	8	872.96
10	Zunis	7	625.63
11	San Mateos	7	366.23
12	N. Uncompaghre	6	359.1
13	Capitans	6	84.8
14	Magdalenas	6	106.8
15	S. Kaibab	4	274.27
16	Pinalenos	4	87.09
17	Navajo	4	6.89
18	Guadalupes	3	25.71
19	Chiricahuas	2	68.13
20	Catalinas	2	29.8
21	Prescotts	2	115.8
22	Hualapais	2	18.25
23	Organs	1	8.57
24	Animas	1	10.58
25	Huachucas	1	12.7
26	Santa Ritas	1	11.9
27	N. Rincon	1	10.08

4. What are some examples of barriers to dispersal of organisms? What are examples of factors that would act as filters to only certain species? Why do you think this study looked at only nonflying mammals?

5. Plot the biodiversity and area values on **Figure 29.4**. The values for the first 10 forest islands and the trend line have been plotted for you.

6. What is the ratio of species/area for the largest park compared with the smallest park? How many species should be in the largest park if the species/area ratio there is the same as the value for the smallest park?

▲ **Figure 29.4** Species numbers and habitat size

Use Figure 29.4 and the trend line to answer the following questions.

7. How many square kilometers of habitat would be required to support 10 species?

    How many square kilometers of habitat would be required to support 5 species?

8. How large is the largest park? How many species does it support? How many species would you expect to find in a park half that size? What percentage reduction is this over the original population?

9. How many square kilometers are required to support four species? How many species would you expect to find in a park half that size? What percent reduction is this over the original population?

10. Based on your answers to Questions 7 and 8, are larger or smaller habitats more prone to species loss due to reduction in habitat size? If you were identifying habitat to protect, would you recommend protecting one 400 km² area or four 100 km² areas? What other factors might you consider? What factors would make a species more deserving of protection? Are all species deserving of equal protection?

11. Optional Google Earth™ activity, Deforestation in the Amazon rain forest. For the KMZ file and questions, go to mygeoscienceplace.com or the Mastering Geography Study Area. Then click on the cover of *Applied Physical Geography: Geosystems in the Laboratory*.

Mastering Geography

Looking for additional lab prep and review materials? Visit the Study Area in Mastering Geography to enhance your geographic literacy and spatial reasoning skills by accessing MapMaster™ interactive maps, videos, *Mobile Field Trips, Project Condor* Quadcopter Videos, *In the News* articles, flashcards, web links, self-study quizzes, and other resources. Contact your professor to find out how you can access Mastering Geography.

Name: _____ Laboratory Section: _____

Date: _____ Score/Grade: _____

Video
Exercise 30
Pre-Lab Video

http://goo.gl/EwTuL0
Scan to view the
Pre-Lab video

LAB EXERCISE

An Introduction to Geographic Information Systems

A **Geographic information system**, known by the familiar **GIS** abbreviation, is a computer-based, data-processing tool for gathering, manipulating, analyzing, and displaying geographic information. Today's sophisticated computer systems allow the integration of geographic information from direct surveys (on-the-ground mapping) and remote sensing in complex ways never before possible. Through a GIS, Earth and human phenomena are analyzed over time. The range of subjects suitable for GIS analysis is limited only by the imagination of the user. If you have used the Internet to get a map or directions, you have used GIS. Today GIS is used to track and predict patterns of disease, help public safety agencies, and plan delivery routes for shipping companies, to name just a few examples. Lab Exercise 30 features an optional online GIS activity at mygeoscienceplace.com and the Mastering Geography Study Area.

Key Terms and Concepts

attribute data
geographic information
   system (GIS)

line
point
polygon

raster
spatial data
vector

KEY**LEARNING**concepts

After completion of this lab, you should be able to:

- *Describe* essential GIS concepts and *identify* the utility of construction of a GIS model.
- *Locate* and *utilize* GIS data on the Internet.
- *Use* an online GIS to *analyze* the risks from Mt. Rainier.

Materials/Sources Needed

Data for analysis
Google Earth

Internet browser access

Lab Exercise and Activities

SECTION 1

Basic GIS Concepts

Modern GIS owes a great deal to the work of Ian McHarg and his search for an ecologically based land-use planning system. He drew on clear acetate sheets to create layers of information. Each layer showed one theme—roads or land cover or existing houses, for example. He stacked multiple sheets to synthesize data and reveal complex relationships. Although he used sheets of plastic to display data (just as you will in the following exercises) rather than a computer screen, the underlying principles are the same. **Figure 30.1** shows an example of how multiple layers are combined.

GISs combine location information and descriptive information for each feature in a layer. The location information tells us where each feature is. These **spatial data** are recorded in a coordinate system such as latitude–longitude. The descriptive information (*attribute data*) tells us about the qualities of each feature. **Attribute data** are stored in a spreadsheet or database and are capable of recording multiple attributes for each location.

Spatial data in a GIS are stored as **points (Figure 30.2a)**, **lines (30.2b)**, or **polygons** (areas, **30.2c**). Features that only need location information are shown as points. Features that have length are stored as arcs, which are made by combining lines. Finally, features with length and width are stored as polygons. This point–line–polygon system is referred to as a **vector data** structure. GIS can also use data stored in another spatial data structure called **raster**. Raster, or grid, cells contain information such as elevation or remote-sensing imagery. Each cell is an intersection of rows and columns, and the attribute it contains may occur anywhere in the cell.

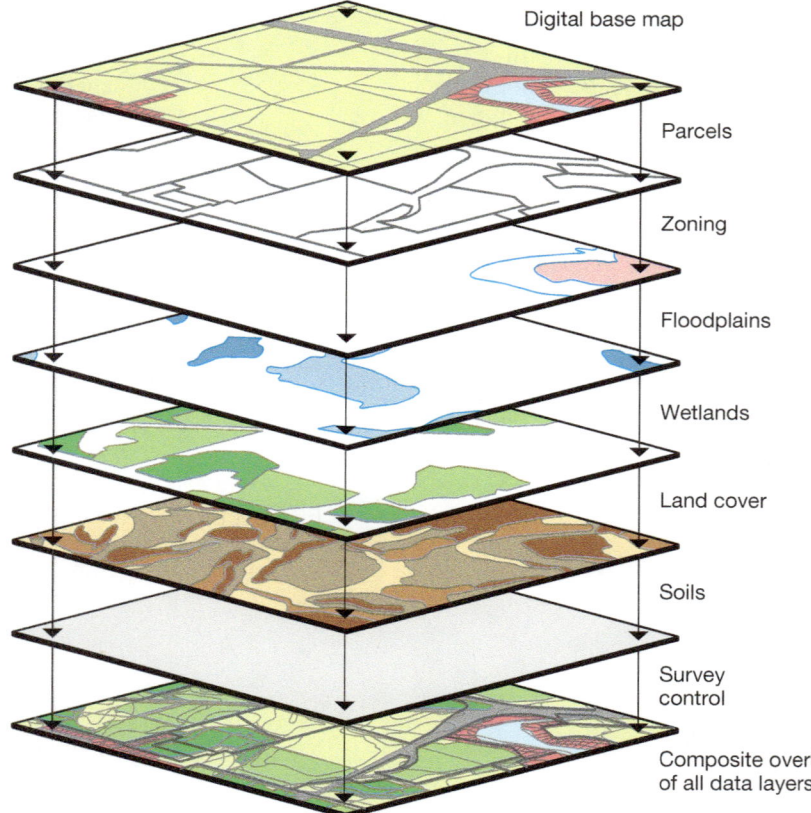

▲ **Figure 30.1** Computer-processed spatial data layered in a geographic information system (GIS) produces a composite overlay for analysis. (After USGS)

▲ Figure 30.2 Examples of point, line, and polygon features: cities (a), highways and rivers (b), and states and counties (c)

By combining layers, a GIS is capable of analyzing patterns and relationships, such as the floodplain or soil layer in Figure 30.1. GIS can create new layers of information with this overlay process as well. A research study may follow specific points or areas through the complex of overlay planes. The utility of a GIS compared with that of a fixed map is the ability to manipulate the variables for analysis and to constantly change the map—the map is alive! Before the advent of computers, an environmental impact analysis required someone to gather data and painstakingly hand-produce overlays of information to determine positive and negative impacts of a project or an event. Today, this layered information is handled by a computer-driven GIS, which assesses the complex interconnections.

1. Give some examples of a GIS that you have used, perhaps without realizing it.

2. If you were a land planner, what questions could you answer by using the composite overlay of data layers in Figure 30.1?

3. Give examples of features in your life that would be symbolized by points, arcs, and polygons.

   Points:

   Arcs:

   Polygons:

SECTION 2

The Internet and GIS

One of the exciting GIS developments in the past few years is the spread of Internet GIS services. You have probably used many of these services without realizing that you were using a GIS. Some examples would include using the Internet to get directions to a location or to show a map. More sophisticated examples of these are Internet Map Servers (IMS) that allow you to pick which layers are displayed, as well as how they are shown. The main advantage of IMS over conventional desktop GIS is that it does not require special software to be installed on the user's computer, nor does using it require special training.

An example of a very sophisticated program that can be used with little training is Ushahidi, which uses a web interface to collect and display crowd-sourced data. It was used very successfully after the devastating earthquake in Haiti in 2010. People in Haiti who needed assistance or supplies were able to send a text message to the Ushahidi team. The text messages were received by a team of volunteers all over the world who then entered the data into a GIS that was used to coordinate rescue efforts. The rescue efforts were also assisted by OpenStreetMap (http://www.openstreetmap.org), another crowd-sourced project. OpenStreetMap is a global public street mapping project where volunteers enter and edit street data. The base data can come from remote-sensing imagery and users' GPS tracks. During the rescue efforts in Haiti, the command center's location was chosen using OpenStreetMap data. For more information about Ushahidi, visit http://www.ushahidi.com. IMS can also use real-time data, such as traffic speed and accident data, to display current road conditions on Google Maps™. Some other examples of IMS are:

> Two online galleries of user-created maps:
> https://maps.google.com/gallery/
> http://www.arcgis.com/home/gallery.html
> An excellent site showing live wildfire information:
> http://www.geomac.gov/

The Internet also has opened up access to free data that was either expensive or completely unavailable just a few years ago. For example, California has created the California Geoportal at http://portal.gis.ca.gov/geoportal, which coordinates GIS data from agencies across the state. This site allows the public to access data easily and eliminates duplication of effort among agencies. A similar site exists at the federal level at http://www.data.gov/geospatial/.

1. Using the Internet, find three websites that use online GIS. Examples might include real estate, online maps or directions, remote-sensing imagery, and wildfire tracking. Record their URLs and descriptions below.

2. Visit three websites that offer free data for downloading that could be used in a GIS. Record their URLs and descriptions below.

SECTION 3

Mt. Rainier Hazards

In this section you will use ArcGIS Online, a web-based GIS. ArcGIS Online is an example of a cloud-based GIS that allows users to perform GIS operations without needing to install programs on a computer.

Mt. Rainier is a large stratovolcano in the Pacific Northwest. It has erupted approximately nine times in the past 10,000 years. The most hazardous products of the eruptions are lahars, volcanic mud flows created by the interaction of snow or ice and lava or pyroclastic debris. For this exercise you will be analyzing lahars with 1000-year and 100-year recurrence intervals. Another hazard is pyroclastic flows of gas, ash, and rock fragments that are over 800°C (1440°F) and can travel at speeds over 160 kmph (100 mph). **Figure 30.3** shows Mt. Rainier and the region of lahar and pyroclastic flow danger. In this exercise, you will produce a volcanic hazards map that will look similar to Figure 30.3.

▲ **Figure 30.3** Mt. Rainier

You are the newly hired emergency preparedness coordinator for the Tri-Counties area. Your area of responsibility is the Washington counties of Thurston, Pierce, and King. Your first order of business is to identify elementary schools that are at risk from the volcanic hazards of Mt. Rainier.

Before you can begin the process, you will need to sign up for a free trial account at www.esri.com/software/arcgis/arcgisonline. The free trial will give you 30 days of access; check with your instructor to find out if your school has an agreement with ESRI that will allow you longer access. After you have created your account for ArcGIS Online, go to www.arcgis.com/home/index.html and sign in with your account information. If you have not used ArcGIS Online, you will want to watch the following tutorial videos: Creating a map; Add features from a file; and Changing symbology. After you have watched the three tutorial videos, you are ready to start making a map.

You will need to download the hazards and schools data for this exercise from http://mygeoscienceplace.com or the Mastering Geography Study Area. Cloud-based GIS is a rapidly changing technology. If ArcGIS Online changes substantially, an updated exercise will be posted to http://mygeoscienceplace.com and the Mastering Geography Study Area.

1. Download the files 100-year_Lahar.zip, 1000-Year_Lahar.zip, Pyroclastic_flow.zip, and Schools.zip to your computer.

2. Sign in to ArcGIS Online and try different base map layers. After you've picked a base map, use the Add tool and Search for Layers at ArcGIS Online. You won't add these layers to the map; this step is just to acquaint you with some of the data sources available. You can limit your search by using key words and specifying the region to search. What are the three most interesting layers that you found?

3. Use the Add tool and Search for Layers at ArcGIS Online and search for topographic maps. You will need a layer that has elevation contour lines, such as the USA Topo Maps layer or the USA Topographic Maps (Mature Support) layer. You can add it to your map or use it as a base map. After you have added an appropriate topographic map, use the Add tool and Add Layer from File for each of the four files you downloaded.

4. Turn off the 100-year lahar and 1000-year lahar layers. Turn each layer on and off so that you can see the extent of each. Modify the symbology of each file so that you can distinguish the pyroclastic flow layer from the 100-year and 1000-year lahar layers.

5. Which layer covers the largest area?

6. You can rearrange the order that the layers draw by dragging and dropping them in the Contents window. In order to best see both lahar layers, which layer should be on the bottom?

7. You can also adjust the transparency of each layer so you can see your data and the base map underneath. What levels of transparency did you use for the pyroclastic flow, 1000-year lahar, and 100-year lahar layers?

8. Use the topographic map and the pyroclastic flow layers to find the relief from summit to the farthest extent of the 100-year lahar flow. How far does the pyroclastic flow extend from the summit? What is the lowest elevation that is covered by potential pyroclastic flow?

9. How far does the 100-year lahar extend from the summit?

10. How far does the 1000-year lahar extend from the summit?

11. What is the name of the school that is closest to the summit and in a lahar area? What is the distance from the summit to the closest school? What is the vertical relief from the summit to the closest school?

12. If lahars can travel at 100 kmph (60 mph), how much time would it take for a flow to reach the school closest to the summit?

13. Identify three schools at risk from a 1000-year lahar but not from a 100-year lahar.

14. Identify three schools at risk from both 1000-year and 100-year lahars.

15. Use the Add tool to search for a Washington counties layer and add that layer to your map. How many schools are at risk from a 1000-year lahar in each county? Which county has the most schools at risk?

16. This exercise involved only elementary schools. What other types of structures would you analyze for hazard potential?

17. Increase the transparency of the lahar layers so you can see the terrain underneath. What terrain features appear to control the flow of the lahars?

18. When you have finished editing your map, use the Share tool to send a link to your instructor.

Data sources:

School data: National Map, http://viewer.nationalmap.gov/viewer/

Mt. Rainier hazards data: Open-File Report 98–428 Volcano Hazards from Mount Rainier, Washington, Revised 1998, http://pubs.usgs.gov/of/1998/0428/

Mastering Geography

Looking for additional lab prep and review materials? Visit the Study Area in Mastering Geography to enhance your geographic literacy and spatial reasoning skills by accessing MapMaster™ interactive maps, videos, *Mobile Field Trips*, *Project Condor* Quadcopter Videos, *In the News* articles, flashcards, web links, self-study quizzes, and other resources. Contact your professor to find out how you can access Mastering Geography.

NOTES

The Lab Exercise in which each term appears is **boldfaced**,
followed by a specific definition related to the usage of the term in *Applied Physical Geography.*

actual evapotranspiration   The actual amount of evaporation and transpiration that occurs; derived in the water balance by subtracting the deficit from potential evapotranspiration (ACTET).

adiabatic   Pertaining to the heating and cooling of a descending or ascending parcel of air through compression and expansion, without any exchange of heat between the parcel and the surrounding environment.

air pressure   Pressure produced by the motion, size, and number of gas molecules and exerted on surfaces in contact with the air. Normal sea level pressure, as measured by the height of a column of mercury (Hg), is expressed as 1013.2 millibars, 760 mm of Hg, or 29.92 inches of Hg. Air pressure can be measured with mercury or aneroid barometers.

albedo   Reflectivity of a surface. The darker a surface, the lower the albedo (the less the surface will reflect the Sun's rays) and the higher the absorption of the Sun's energy.

Alber's projection   Alber's equal-area conic projection, used as a base for U.S. quadrangle maps. Two standard parallels are used for the conterminous United States—29.5°N and 45.5° N latitudes. The standard parallels are shifted for conic projections of Alaska (55°N and 65°N) and Hawai'i (8°N and 18°N).

alluvial fan   A fan-shaped fluvial landform at the mouth of a canyon, particularly noticeable in arid landscapes where streams are intermittent.

alluvial terraces   Level areas that appear as topographic steps above a stream, created by the stream as it scours with renewed downcutting into its floodplain; composed of unconsolidated alluvium (see *alluvium*).

alluvium   General descriptive term for clay, silt, and sand transported by running water and deposited in sorted or semi-sorted sediment on a floodplain, delta, or in a stream bed.

alpine glacier   A glacier confined in a mountain valley or walled basin, consisting of three subtypes: valley glacier (within a valley), piedmont glacier (coalesced at the base of a mountain, spreading freely over nearby lowlands), and outlet glacier (flowing outward from a continental glacier).

altimetry   The measurement of altitude using air pressure. A *pressure altimeter* is an instrument that measures altitude based on a strict relationship between air pressure and altitude. Air pressure is measured within the altimeter by an aneroid barometer capsule with the instrument graduated in increments of altitude.

altitude   The angular distance between the horizon (a horizontal) and the Sun (or any point).

analemma   A convenient device to track the passage of the Sun, the Sun's declination, and the positive and negative equation of time throughout the year.

analog   Using a different method to portray some type of information. Relating to GIS, the use of manually prepared graphics (e.g., transparencies) to depict spatial information.

aneroid barometer   A device to measure air pressure using a partially emptied, sealed cell (see *air pressure*).

anticline   The upfolds in a folded landscape, often creating ridges between the downfolded *synclines*.

apparent temperature   The temperature subjectively perceived by each individual, also known as sensible temperature.

arctic tundra   A biome in the northernmost portions of North America and northern Europe and Russia, featuring low, ground-level herbaceous plants as well as some woody plants.

arêtes   A sharp ridge that divides two cirque basins. Derived from "knife edge" in French, these form sawtooth and serrated ridges in glaciated mountains.

asthenosphere   The region of the upper mantle known as the plastic layer; the least rigid portion of Earth's interior; shatters if struck yet flows under extreme heat and pressure.

attribute table   Nonspatial information about a feature being mapped in a GIS (e.g., age, use, type, value), as contrasted to *spatial data.*

available water   The portion of capillary water that is accessible to plant roots; usable water held in soil moisture storage.

azimuth   A compass direction expressed as an arc area clockwise from the north through 360°.

backswamp   A low-lying, swampy area of a floodplain; adjacent to a river, with the river's natural levees on one side and the sides of the valley on the other (see *yazoo tributary*).

bajada   A continuous apron of coalesced alluvial fans, formed along the base of mountains in arid climates; presents a gently rolling surface from fan to fan. (See Alluvial fan.)

barometer   A device to measure air pressure; either mercury or aneroid type (see *air pressure*).

barrier spit   A depositional form that develops when transported sand in a barrier beach or island is deposited in long ridges that are attached at one end to the mainland and partially cross the mouth of a bay.

basin and range   A region of dry climates, few permanent streams, and interior drainage patterns in the western United States; a faulted landscape composed of a sequence of horsts and grabens.

bay barrier An extensive sand spit that encloses a bay, cutting it off completely from the ocean and forming a lagoon; produced by littoral drift and wave action; sometimes referred to as a *baymouth bar*.

beach The portion of the coastline where an accumulation of sediment is in motion.

biodiversity A principle of ecology and biogeography which says that the more diverse the species population in an ecosystem (in number of species, quantity of members in each species, and genetic content), the more risk is spread over the entire community, which results in greater overall stability, greater productivity, and increased use of nutrients, as compared to a monoculture of little or no diversity.

biomass The total mass of living organisms on Earth or per unit area of a landscape; also, the weight of the living organisms in an ecosystem.

biome A large terrestrial ecosystem characterized by specific plant communities and formations; usually named after the predominant vegetation in the region.

boreal forest See *northern needleleaf forest*.

capillary water Soil moisture, most of which is accessible to plant roots; held in the soil by surface tension and cohesive forces between water and soil (see also *available water, field capacity*, and *wilting point*).

cardinal points The four principal compass points: north, south, east, and west.

cartography The making of maps and charts; a specialized science and art that blends aspects of geography, engineering, mathematics, graphics, computer science, and artistic specialties.

cinder cone A volcanic landform of pyroclastics and scoria, usually small and cone shaped and generally not more than 450 m (1500 ft) in height, with a truncated top.

circle of illumination The division between lightness and darkness on Earth; a day-night great circle.

cirque A scooped-out, amphitheater-shaped basin at the head of an alpine glacier valley; an erosional landform.

classification The process of ordering or grouping data or phenomena in related classes; results in a regular distribution of information; a taxonomy.

climate The consistent, long-term behavior of weather over time, including its variability; in contrast to *weather*, which is the condition of the atmosphere at any given place and time.

climate change A large and lasting change in the statistical distribution of weather patterns. The time scale of these changes may range from decades to millions of years. The changes include average weather conditions, or in the distribution of weather around the average conditions (i.e., more or fewer extreme weather events).

climatic regions Areas of similar climate, which contain characteristic regional weather and air mass patterns.

climatology A scientific study of climate and climatic patterns and the consistent behavior of weather and weather variability and extremes over time in one place or region and the effects climate and climate change have on human society and culture.

climograph A graph that plots daily, monthly, or annual temperature and precipitation values for a selected station; may also include additional weather information.

col Formed by two headward-eroding cirques that reduce an arête (ridge crest) to form a high pass.

community A convenient biotic subdivision within an ecosystem; formed by interacting populations of animals and plants in an area.

compass points The spots marking the 32 directional points on a compass.

composite volcano A volcano formed by a sequence of explosive volcanic eruptions; steep-sided, conical in shape; sometimes referred to as a stratovolcano, although *composite* is the preferred term (compare *shield volcano*).

conic A class of map projections based on the geometric shape of a cone.

continental drift A proposal by Wegener in 1912 stating that Earth's landmasses have migrated over the past 200 million years from a supercontinent he called Pangaea to the present configuration; a widely accepted concept today (see *plate tectonics*).

continental glacier A continuous mass of unconfined ice, covering at least 50,000 km² (19,500 mi²); most extensive as ice sheets covering Greenland and Antarctica.

continentality A qualitative designation applied to stations that lack the temperature-moderating effects of the sea and that exhibit a greater range between minimum and maximum temperatures, both daily and annually.

contour interval The vertical distance between adjacent contour lines (see *contour lines*).

contour lines Isolines on a topographic map that connect all points at the same elevation relative to a reference elevation called the vertical datum. The spaces between contour lines portray slope and are known as *contour intervals*.

convergent boundary A region where two tectonic plates converge, causing compression or subduction of one or both plates involved.

Coordinated Universal Time (UTC) The official reference time in all countries, formerly known as Greenwich Mean Time outside the United Kingdom; now measured by primary standard atomic clocks whose time calculations are collected at the Paris, France, International Bureau of Weights and Measures (BIPM). UTC assumes a prime meridian (0° longitude) passing through Greenwich, London, England.

core The deepest inner portion of Earth, representing one-third of its entire mass; differentiated into two zones—a solid iron inner core surrounded by a dense, molten, fluid metallic-iron outer core.

crater A circular surface depression formed by volcanism; built by accumulation, collapse, or explosion; usually located at a volcanic vent or pipe; can be at the summit or on the flank of a volcano.

crust Earth's outer shell of crystalline surface rock, ranging from 5 to 60 km (3 to 38 mi) in thickness from oceanic crust to mountain ranges. Average density of continental crust is crust is 2.7 g per cm^3, whereas oceanic crust is 3.0 g per cm^3.

cylindrical A class of map projections based on the geometric shape of a cylinder.

day length The duration of exposure to insolation, varying during the year depending on latitude; an important aspect of seasonality.

daylight saving time A period in the Northern Hemisphere during which time is set ahead one hour in the spring and set back one hour in the fall. Only Hawai'i, Arizona, and Saskatchewan exempt themselves. In Europe, the last Sundays in March and September generally are used.

declination The latitude that receives direct overhead (perpendicular) insolation on a particular day; migrates annually through the 47° of latitude between the tropics.

deficit In a water balance, the amount of unmet, or unsatisfied, potential evapotranspiration (DEFIC).

denudation A general term that refers to all processes that cause degradation of the landscape: weathering, mass movement, erosion, and transportation; at the heart of exogenic processes.

desert biomes Arid landscapes of uniquely adapted dry-climate plants and animals.

dew-point temperature The temperature at which a given mass of air becomes saturated, holding all the water it can hold. Any further cooling or addition of water vapor results in active condensation.

dike An intrusive igneous feature formed when magma intrudes across layers of existing rock.

discharge The measured volume of flow in a river that passes by a given cross section of a stream in a given unit of time; expressed in cubic meters per second or cubic feet per second.

divergent boundary A region where upwelling material from the mantle forms new crust and tectonic plates spread apart in a constructive process.

drainage basins The basic spatial geomorphic unit of a river system; distinguished from a neighboring basin by ridges and highlands that form divides delimiting the catchment area of the drainage basin, or its watershed.

drainage patterns A geometric arrangement of streams in a region; determined by slope, differing rock resistance to weathering and erosion, climatic and hydrologic variability, and structural controls of the landscape.

drumlin A depositional landform related to continental glaciation; an elongated hill created from till embedded in the bottom of the glacial ice that is plastered, layer upon layer, in a growing mound.

dry adiabatic rate (DAR) The rate at which a parcel of air that is less than saturated cools (if ascending) or heats (if descending); a rate of 10C° per 1000 m (5.5F° per 1000 ft) (see also *adiabatic*).

ecosystem A self-regulating association of living plants, animals, and their nonliving physical and chemical environment.

effusive eruption An eruption characterized by low-viscosity, basaltic magma, with characteristic low-gas content readily escaping. Lava pours forth onto the surface with relatively small explosions and little tephra; tends to form shield volcanoes.

elastic-rebound theory A concept describing the faulting process in Earth's crust, in which the two sides of a fault appear locked despite the motion of adjoining pieces of crust; with accumulating strain, they rupture suddenly, snapping to new positions relative to each other, generating an earthquake.

empirical classification A classification based on weather statistics or other data; used to determine general climate categories.

environmental lapse rate The actual lapse rate in the lower atmosphere at any particular time under local weather conditions; may deviate above or below the average normal lapse rate of 6.4C° per 1000 m (3.5F° per 1000 ft).

equal area A trait of a map projection; indicates the equivalence of all areas on the surface of the map, although shape is distorted.

equatorial and tropical rain forest A lush biome of tall broadleaf evergreen trees and diverse plants and animals. The dense canopy of leaves is usually arranged in three levels.

equilibrium line The area of a glacier where accumulation (gain) and ablation (loss) are balanced.

equinox March 20–21 and September 22–23, when the circle of illumination divides Earth through the poles, and all places have days and nights of equal length.

esker A sinuously curving, narrow deposit of coarse gravel that forms along a meltwater stream channel, developing in a tunnel beneath the glacier. A retreating glacier leaves such ridges behind in a pattern that usually parallels the path of the glacier; may appear branching.

evaporation The movement of free water molecules away from a wet surface into air that is less than saturated; the phase change of water to water vapor; vaporization below the boiling point of water.

evapotranspiration The merging of evaporation and transpiration water loss into one term (see *potential evapotranspiration* and *actual evapotranspiration*).

explosive eruption A violent and unpredictable volcanic eruption, the result of magma that is thicker (more viscous), stickier, and higher in gas and silica content than that of an effusive eruption; tends to form blockages within a volcano; produces composite volcanic landforms (see *composite volcano*; compare *effusive eruption*).

extratropical cyclone See *midlatitude cyclone*.

extrusive igneous rock A rock, such as basalt, that solidifies and crystallizes from a molten state as it extrudes onto the surface.

felsic A type of igneous rock, such as granite, that is high in silicate minerals, such as feldspar and quartz (pure silica), and has a low melting point. Their name is derived from *fel*dspar and *si*lica. Rocks formed from felsic minerals generally are lighter in color and less dense than those from mafic minerals.

field capacity Water held in the soil by hydrogen bonding against the pull of gravity, remaining after water drains from the larger pore spaces, or storage capacity; the available water for plants. Field capacity is specific to each soil type and is an amount that can be determined through soil surveys.

flood A high water level that overflows the natural riverbank along any portion of a stream.

flood basalts An accumulation of horizontal flows formed when lava spreads out from elongated fissures onto the surface in extensive sheets; associated with effusive eruptions.

floodplain A low-lying area near a stream channel that is subject to recurrent flooding; alluvial deposits generally mask underlying rock.

fluvial Stream-related processes; from the Latin *fluvius*, for "river" or "running water."

formation class The dominant vegetation type extending across a region.

gage height The height of the water surface above the gage datum (zero point).

genetic classification A type of classification that uses causative factors to determine climatic regions; for example, an analysis of the effect of interacting air masses on climate.

geographic grid A network of intersecting lines running north–south (through the poles) and east–west (parallel to the equator) that provides a system of global location (see *latitude* and *longitude*).

geographic index number The identification number on a topographic quadrangle, derived from the latitude and longitude of the lower-right corner of the map.

Geographic Information Systems (GIS) A computer-based methodology that merges computer cartography with data management and allows the collection, manipulation, and analysis of geographic/spatial information.

geographic North (or South) Pole The geographic North (or South) Pole—the point at which all meridians converge—giving a reference point for true north (or south); also known as *true north* (or *true south*).

geography (Preface) The science that studies the interdependence among geographic areas, natural systems, processes, society, and cultural activities over space—a spatial science. The five themes of geographic education are location, place, movement, regions, and human–Earth relationships.

geomorphology The science that analyzes and describes the origin, evolution, form, classification, and spatial distribution of landforms.

glacier A large mass of perennial ice resting on land or floating shelflike in the sea adjacent to land; formed from the accumulation and recrystallization of snow, which then flows slowly under the pressure of its own weight and the pull of gravity.

gnomonic projection A planar map projection on which all straight lines are great circle routes.

great circle Any circle of circumference drawn on a globe with its center coinciding with the center of the globe. An infinite number of great circles can be drawn, but only one parallel is a great circle—the equator.

Greenwich Mean Time (GMT) Former world standard time, now known as Coordinated Universal Time (UTC) (see *Coordinated Universal Time*).

habitat The physical location in which an organism is biologically suited to live. Most species have specific habitat parameters, or limits.

hair hygrometer An instrument for the measurement of relative humidity; based on the principle that human hair will change as much as 4% in length between 0% and 100% relative humidity.

heat index An index that indicates the human body's reaction to air temperature and water vapor. The level of humidity in the air affects our natural ability to cool through evaporation from skin.

horn A pyramidal, sharp-pointed peak that results when several cirque glaciers gouge an individual mountain summit from all sides.

horst and graben Landscape features in the Basin and Range Province. The term *horst* applies to upward-faulted blocks; *graben* refers to downward-faulted blocks.

hot spots Individual points of upwelling material originating in the asthenosphere, not necessarily associated with spreading centers, although Iceland is astride a mid-ocean ridge. They tend to remain fixed relative to migrating plates; some 50 to 100 are identified worldwide, exemplified by Yellowstone National Park and Hawai'i (see *plumes*).

humidity Water vapor content of the air. The capacity of the air to hold water vapor is mostly a function of temperature.

humus A mixture of organic debris in the soil, worked by consumers and decomposers in the humification process; characteristically formed from plant and animal litter deposited at the surface.

hurricane A tropical cyclone that is fully organized and intensified in inward-spiraling rainbands; ranges from 160 to 960 km (100 to 600 mi) in diameter, with wind speeds in excess of 119 kmph (65 knots, or 74 mph); a name used specifically in the Atlantic and eastern Pacific (compare *typhoon*).

hydrology The science that studies the flow of water, ice, and water vapor from place to place, as water flows through the atmosphere, across the land, where it is also stored as ice, and within groundwater; includes interactions with human activity.

ice core A column of glacial ice used to study past climate.

igneous rock One of the three basic rock types; it has solidified and crystallized from a hot molten state (either magma or lava) (compare *metamorphic rock* and *sedimentary rock*).

insolation Solar radiation that is intercepted by Earth.

interior drainage In regions where rivers do not flow into the ocean, the outflow is through evaporation or subsurface gravitational flow. Portions of Africa, Asia, Australia, and the western United States have such drainage.

internal drainage Drainage in regions where rivers do not flow into the ocean, and the outflow is through evaporation or subsurface gravitational flow. Portions of Africa, Asia, Australia, and the western United States have such drainage.

International Date Line The 180° meridian; an important corollary to the prime meridian on the opposite side of the planet; established by the treaty of 1884 to mark the place where each day officially begins.

intrusive igneous rock A rock, such as granite, that solidifies and crystallizes from a molten state as it intrudes into crustal rocks, cooling and hardening below the surface.

island biogeography The study of island communities, which are special places for study because of their spatial isolation and the relatively small number of species present. Islands resemble natural experiments because the impact of individual factors, such as civilization, can be more easily assessed on islands than over larger continental areas.

isobar An isoline connecting all points of equal atmospheric pressure.

isogon An isoline connecting points of equal magnetic declination (see *magnetic declination*).

isogonal map A map showing isogons (see *isogon*).

isoline A line on a map that connects all points of equal value.

isotherm An isoline that connects all points of equal air temperature.

isotope analysis A technique for long-term climatic reconstruction that uses the atomic structure of chemical elements—specifically the relative amounts of their isotopes—to identify the chemical composition of past oceans and ice masses.

kame A small hill of poorly sorted sand and gravel that accumulates in crevasses or in ice-caused indentations in the surface; may also be deposited by water in ice-contact deposits.

karst topography Distinctive topography formed in a region of chemically weathered limestone with poorly developed surface drainage and solution features that appear pitted and bumpy; originally named after the Krš Plateau of Yugoslavia.

kettle A geographic feature that forms when an isolated block of ice persists in a ground moraine, an outwash plain, or valley floor after a glacier retreats; as the block finally melts, it leaves behind a steep-sided hole that may frequently fill with water.

kinetic energy The energy of motion in a body; derived from the vibration of the body's own movement and stated as temperature.

lagoon A portion of coastal seawater that is virtually cut off from the ocean by a bay barrier or barrier beach; also, the water surrounded and enclosed by an atoll.

landfall The location along a coast where a storm moves onshore.

land-water heating differences The differences in the way land and water heat, as a result of contrasts in transmission, evaporation, mixing, and specific heat capacities. Land surfaces heat and cool faster than water and are characterized as having aspects of continentality, whereas water is regarded as producing a moderating marine influence.

latent heat The heat energy absorbed or released in the phase change of water from one state to another. The energy is absorbed or released in each phase change from one state to another. Heat energy is absorbed as the latent heat of melting, vaporization, or evaporation. Heat energy is released as the latent heat of condensation and freezing (or fusion).

latitude The angular distance measured north or south of the equator from a point at the center of Earth. A line connecting all points of the same latitudinal angle is called a *parallel*.

lava Magma that issues from volcanic activity onto the surface; the extrusive rock that results when magma solidifies (see *magma*).

life zone An altitudinal zonation of plants and animals that form distinctive communities. Each life zone possesses its own temperature and precipitation relationships.

lifting condensation level The height at which an air parcel will reach 100% relative humidity (RH) when it is cooled by dry adiabatic lifting.

line A GIS vector feature with one dimension-length.

littoral zone A specific coastal environment that is the region between the high-water line during storms and the depth at which storm waves cannot move sea-floor sediments.

loam A soil that is a mixture of sand, silt, and clay in almost equal proportions, with no one texture dominant; an ideal agricultural soil.

local Sun time Time based on actual longitudinal distance and the prime meridian; also referred to as Sun time or solar time.

longitude The angular distance measured east or west of a prime meridian from a point at the center of Earth. A line connecting all points of the same longitude is called a meridian.

mafic A type of igneous rock, such as basalt, that is derived from *magnesium* and *ferrum* (the Latin word for iron). Mafic rocks are lower in silica, higher in magnesium and iron, have higher melting points, are darker in color, and have greater density than those from felsic minerals.

magnetic declination The horizontal angle between geographic (true) north and magnetic north.

magnetic North (or South) Pole The point on Earth's surface indicated by the north-seeking point of a magnetic compass needle.

mantle An area within the planet representing about 80% of Earth's total volume, with densities increasing with depth and averaging 4.5 g per cm³; occurs above the core and below the crust; is rich in iron and magnesium oxides and silicates.

map, map projection The reduction of a spherical globe onto a flat surface in some orderly and systematic realignment of the latitude and longitude grid.

march of the seasons The annual passage of Earth as it revolves around the Sun, producing seasonal changes in daylength and Sun altitude.

marine Descriptive of stations that are dominated by the moderating effects of the ocean and that exhibit a smaller minimum and maximum temperature range than continental stations (see *land-water heating differences*).

Mediterranean shrubland A major biome dominated by shrub formations; occurs in Mediterranean dry summer climates and is characterized by sclerophyllous scrub and short, stunted, tough forests.

Mercator projection A true-shape cylindrical projection on which straight lines are true compass directions (lines of constant bearing)—rhumb lines.

mercury barometer A device that measures air pressure with a column of mercury in a tube that is inserted in a vessel of mercury. The surrounding air exerts pressure on the mercury in the vessel (see *air pressure*).

meridian See *longitude*.

mesopause The upper boundary of the mesosphere.

mesosphere The upper region of the homosphere from 50 to 80 km (30 to 50 mi) above the ground; designated by temperature criteria and very low pressures, ranging from 0.1 to 0.001 mb.
The top of the mesosphere, as determined by temperature ($-90°C$), is the mesopause.

metamorphic rocks One of the three basic rock types; it is existing igneous and sedimentary rock that has undergone profound physical and chemical changes under increased pressure and temperature. Constituent mineral structures may exhibit foliated or nonfoliated textures (compare *igneous rock* and *sedimentary rock*).

meteorology The scientific study of the atmosphere, including a study of the atmo-sphere's physical characteristics and motions, related chemical, physical, and geological processes, the complex linkages of atmospheric systems, and weather forecasting.

mid-ocean ridges Submarine mountain ranges that extend more than 65,000 km (40,000 mi) worldwide and average more than 1000 km (620 mi) in width; centered along sea-floor spreading centers as the direct result of upwelling areas of heat in the upper mantle.

midlatitude broadleaf and mixed forest A biome in moist continental climates in areas of warm-to-hot summers and cool-to-cold winters; includes several distinct communities. Relatively lush stands of broadleaf forests trend northward into needleleaf evergreen stands.

midlatitude grasslands The major biome most modified by human activity; so named because of the predominance of grass-like plants, although deciduous broadleafs appear along streams and other limited sites; location of the world's breadbaskets of grain and livestock production.

mineral An element or combination of elements that forms an inorganic natural compound; described by a specific formula and crystal structure.

moist adiabatic rate (MAR) The rate at which a parcel of saturated air cools in ascent; a rate of 6C° per 1000 m (3.3F° per 1000 ft). This rate may vary, with moisture content and temperature, from 4C° to 10C° per 1000 m (2F° to 6F° per 1000 ft). Also referred to as wet adiabatic rate (WAR) (see *adiabatic*).

montane forests Needleleaf forests associated with mountain elevations (see *northern needleleaf forest*).

moraine Marginal glacial deposits of unsorted and unstratified material, producing a variety of depositional landforms.

natural levees Long, low ridges that occur on either side of a river in a developed floodplain; sedimentary (coarse gravels and sand) depositional by-products of river-flooding episodes.

niche The basic function, or occupation, of a life-form within a given community; the way an organism obtains its food, air, and water.

normal lapse rate The average rate of temperature decrease with increasing altitude in the lower atmosphere; an average value of 6.4C° per km, or 1000 m (3.5F° per 1000 ft); a rate that exists principally during daytime conditions.

northern needleleaf forest Forests of pine, spruce, fir, and larch, stretching from the east coast of Canada westward to Alaska and continuing from Siberia westward across the entire extent of Russia to the European Plain; called the taiga (a Russian word) or the boreal forest; principally in the D climates. Includes montane forests.

orographic lifting The uplifting of migrating air masses in response to the physical presence of a mountain, a topographic barrier. The lifted air cools adiabatically as it moves upslope; may form clouds and produce increased precipitation (see *rain shadow*).

outwash plain Glacio-fluvial deposits of stratified drift of meltwater-fed, braided, and overloaded streams with debris beyond a glacier's morainal deposits.

oxbow lake A lake that was formerly part of the channel of a meandering stream; isolated when a stream eroded its outer bank, forming a cutoff through the neck of a looping meander.

paleoclimatology The science that studies climates and the causes of variations in climate of past ages throughout historic and geologic time.

Pangaea The supercontinent formed by the collision of all continental masses approximately 225 million years ago; named by Wegener in 1912 in his continental drift theory.

parallel See *latitude*.

paternoster lake One of a series of small, circular, stair-stepped lakes formed in individual rock basins aligned down the course of a glaciated valley; named because they look like a string of rosary (religious) beads.

paternoster lakes One of a series of small, circular, stair-stepped lakes formed in individual rock basins aligned down the course of a glaciated valley; named because they look like a string of rosary (religious) beads.

ped The smallest natural lump or cluster of particles in soil.

pedon A soil profile extending from the surface to the lowest extent of plant roots or to the depth where regolith or bedrock is encountered; imagined as a hexagonal column; the basic soil sampling unit.

permeability The ability of water to flow through soil or rock; a function of the texture and structure of the medium.

physical geography (Preface) A science that studies the spatial aspects of the physical elements and processes that make up the environment: energy, air, water, weather, climate, landforms, soils, animals, plants, and Earth.

planar A class of map projections based on the geometric shape of a plane. Also known as azimuthal.

planimetric map A basic map showing the horizontal position of boundaries, land-use activities, and political, economic, and social outlines.

plate tectonics The conceptual model that encompasses continental drift, sea-floor spreading, and related aspects of crustal movement; widely accepted as the foundation of crustal tectonic processes.

playa An area of salt crust left behind by evaporation on a desert floor, usually in the middle of a desert or semiarid bolson or valley; intermittently wet and dry.

plumes Columns of heated rock rising through the asthenosphere or mantle, creating hot spots in the migrating overlying lithospheric plates (see *hot spots*).

point GIS vector feature that consists of a location.

polygon GIS vector feature of an area.

polypedon The identifiable soil in an area, with distinctive characteristics differentiating it from surrounding polypedons that form the basic mapping unit; composed of many pedons (see *pedon*).

porosity The total volume of available pore space in soil; a result of the texture and structure of the soil.

potential evapotranspiration The amount of moisture that would evaporate and transpire if adequate moisture were available; the amount lost under optimum moisture conditions—that is, the moisture demand (POTET).

precipitation Rain, snow, sleet, and hail—the moisture supply (PRECIP).

pressure gradient 7 The change in atmospheric pressure horizontally across Earth's surface; measured along a line at right angles to the isobars.

prime meridian An arbitrary meridian designated as 0° longitude; the point from which longitude is measured east or west; agreed on by the nations of the world in an 1884 treaty.

profile view See *topographic profile*.

proxy method A method that finds information about past environments that indicates changes in climate, such as isotope analysis or tree ring dating; also called a *climate proxy*.

pyroclastics An explosively ejected rock fragment launched by a volcanic eruption; sometimes described by the more general term *tephra*.

quadrant compass bearing A compass direction expressed as an arc clockwise or counterclockwise (toward the east or toward the west—whichever is closer) through 90° from the north or the south.

rain shadow The area on the leeward slopes of a mountain range; in the shadow of the mountains, where precipitation receipt is greatly reduced compared to windward slopes (see *orographic lifting*).

raster In GIS, a data structure based on a grid of (usually) square and rectangular "cells."

rating curve   A curve that relates stream discharge to gage height in showing the direct relationship between stream discharge and its width, depth (gage height), and velocity.

recurrence intervals   The average time period within which a given event will be equaled or exceeded once; also called *return period*.

regime   The characteristic seasonal variation in streamflow, reflecting climatic conditions—a stream's annual "personality."

regolith   Partially weathered rock overlying bedrock, whether residual or transported.

relative humidity   A term that reflects the ratio of water vapor actually in the air (content) compared to the maximum water vapor the air is able to hold (capacity) at that temperature; expressed as a percentage.

relief   Elevation differences in a local landscape as an expression of unevenness, height, and slope variation.

revolution   The annual orbital movement of Earth about the Sun; determines year length and the length of seasons.

rhumb line   A line of constant compass direction, or constant bearing, which crosses all meridians at the same angle. A portion of a great circle.

rock   An assemblage of minerals bound together, or sometimes a mass of a single mineral.

rock cycle   A model that represents the interrelationships among the three rock-forming processes: igneous, sedimentary, and metamorphic; shows how each can be transformed into another rock type.

rotation   The turning of Earth on its axis; averages 24 hours in duration; determines day–night relationships.

saturated   Descriptive of air that is holding all the water vapor that it can hold at a given temperature.

scale   The ratio of the distance on a map to the distance in the real world; expressed as a representative fraction, graphic scale, or written scale.

sea cliff   Cliffs formed by the undercutting action of the sea.

sea-floor spreading   As proposed by Hess and Dietz, the mechanism driving the movement of the continents; associated with upwelling flows of magma along the worldwide system of mid-ocean ridges.

sedimentary rocks   One of the three basic rock types; it is formed from the compaction, cementation, and hardening of sediments derived from other rocks (compare *igneous rock* and *metamorphic rock*).

seismic waves   The shock waves sent through the planet by an earthquake or underground nuclear test. Transmission varies according to temperature and the density of various layers within the planet.

sensible heat   Heat that can be measured with a thermometer; a measure of the concentration of kinetic energy from molecular motion (see *apparent temperature*).

shield volcano   A symmetrical mountain landform built from effusive eruptions (low-viscosity magma); gently sloped and gradually rising from the surrounding landscape to a summit crater; typical of the Hawaiian Islands (compare *composite volcano* and *effusive eruption*).

sinkhole   A nearly circular depression created by the weathering of karst landscapes; also known as a doline in traditional studies; may collapse through the roof of an underground space (see *karst topography*).

sling psychrometer   A device for the measurement of relative humidity using two thermometers—a dry bulb and a wet bulb—mounted side-by-side.

small circle   Circles on a globe's surface that do not share Earth's center; for example, all parallels other than the equator.

soil   A dynamic natural body made up of fine materials covering Earth's surface in which plants grow, composed of both mineral and organic matter.

soil horizon   The various layers exposed in a pedon; roughly parallel to the surface and identified as O, A, E, B, C, and R (bedrock).

soil pH (acidity-alkalinity)   The relative abundance of free hydrogen ions (H^+) in a solution. The pH scale is logarithmic: Each whole number represents a 10-fold change. A pH of 7.0 is neutral, and values less than 7.0 are increasingly acidic, while values greater than 7.0 are increasingly basic or alkaline.

soil moisture storage; recharge; utilization; change   The retention of moisture within soil; represents a savings account that can accept deposits (*soil moisture recharge*) or experiences withdrawals (*soil moisture utilization*) as conditions change.

soil profile   See *pedon*.

soil science   An interdisciplinary science that studies soil and involves physics, chemistry, biology, mineralogy, hydrology, taxonomy, climatology, and cartography.

soil structure   The size and shape of the aggregates of particles in the soil.

soil-water budget   An accounting system for soil moisture that uses inputs of precipitation and outputs of evapotranspiration and gravitational water.

solstice   June 20–21 and December 21–22, when the Sun's declination is at the Tropics of Cancer or Capricorn, respectively.

solum   A true soil profile in the pedon; ideally, a combination of O, A, E, and B horizons. (See Pedon.)

spatial data   In a GIS, the data that are concerned with the location of a feature, as contrasted with *attribute data*.

specific heat   The increase of temperature in a material when energy is absorbed. Because water requires far more heat to raise its temperature than does a comparable volume of land, water is said to have a higher specific heat.

specific humidity The mass of water vapor (in grams) per unit mass of air (in kilograms) at any specified temperature. The maximum mass of water vapor that a kilogram of air can hold at any specified temperature is termed its maximum specific humidity.

stability The condition of a parcel of air relative to whether it remains where it is or changes its initial position. The parcel is stable if it resists displacement upward and unstable if it continues to rise; relates to adiabatic and lapse rate processes.

standard atmosphere An agreed-upon model of Earth's atmosphere, with set values for the vertical distribution of temperature, pressure, and density from the surface to approximately 80 km (50 mi).

standard line The line of tangency on a map projection along which the projection surface (cylinder or cone) contacts the globe; along this line the map maintains accuracy, free of distortion.

storm surge Large quantities of seawater pushed inland by the strong winds associated with a tropical cyclone.

stratopause The upper boundary of the stratosphere.

stratosphere The portion of the homosphere that ranges from 20 to 50 km (12.5 to 30 mi) above Earth's surface, with temperatures ranging from $-57°C$ ($-70°F$) at the tropopause to $0°C$ ($+32°F$) at the stratopause. The functional ozonosphere is within the stratosphere.

subduction zone An area where two crustal plates collide and the denser oceanic crust dives beneath the less dense continental plate, forming deep oceanic trenches and seismically active regions.

subsolar point The only point receiving perpendicular insolation at a given moment—that is, where the Sun is directly overhead.

surplus The amount of moisture that exceeds potential evapotranspiration; moisture oversupply when soil moisture storage is at field capacity.

syncline In a landscape formed by folding, the downfold or trough often creating valleys between the upfolded *anticlines*.

synoptic map A weather map that shows a moment in time over a large area.

tarn A small mountain lake that occupies a basin in a glacial trough or cirque.

temperate rain forest A major biome of lush forests at middle and high latitudes; occurs along narrow margins of the Pacific Northwest in North America, among other locations; a mixture of broadleaf and needleleaf trees and thick undergrowth, including the tallest trees in the world.

temperature A measure of sensible heat energy present in the atmosphere and other media; indicates the average kinetic energy of individual molecules within the atmosphere.

terrestrial ecosystem A self-regulating association characterized by specific plant formations; usually named for the predominant vegetation and known as a biome when large and stable.

thermal equator The isoline on an isothermal map that connects all points of highest mean temperature.

thermopause The upper boundary of the thermosphere.

thermosphere A region of the heterosphere extending from 80 to 480 km (50 to 300 mi), the thermopause, in altitude; contains the functional ionosphere layer.

till plains Large, relatively flat plains composed of unsorted glacial deposits behind a terminal or end moraine. Low-rolling relief and interrupted, unclear drainage patterns are characteristic.

tombolo A landform created when coastal sand deposits connect the shoreline with an offshore island outcrop or sea stack.

topographic map A map that portrays physical relief through the use of elevation contour lines that connect all points at the same elevation above or below a vertical datum, such as mean sea level.

topographic profile A graphic representation of graduated elevations along a line segment drawn on a map—a *profile view*—that is an important method of analysis for a topographic map.

topography The undulations and configurations that give Earth's surface its texture; the heights and depths of local relief, including both natural and human-made features.

township and range A system that the Public Lands Survey System delineated in 1785 that called for the establishment of a grid system based on an initial survey point; most of the United States is subdivided into 36 mi^2 townships, 1 mi^2 sections, and 160 acre homesteads (quarter section).

transform boundary A region where tectonic plates slide past one another, usually at right angles to a seafloor spreading center.

transform faults An elongate zone along which faulting occurs between mid-ocean ridges; produces a relative horizontal motion with no new crust formed or destroyed; strike-slip motion, either left or right lateral.

transpiration The movement of water vapor out through the pores in leaves drawn by their roots from the soil moisture storage.

tropical cyclone A cyclonic circulation originating in the tropics, with winds between 30 and 64 knots (39 and 73 mph); characterized by closed isobars, circular organization, and heavy rains (see *hurricane* and *typhoon*).

tropical savanna A major biome that contains large expanses of grassland interrupted by trees and shrubs; a transitional area between the humid rain forests and tropical seasonal forests and the drier, semiarid tropical steppes and deserts.

tropical seasonal forest and scrub A variable biome on the margins of the rain forests, occupying regions of lesser and more erratic rainfall; the site of transitional communities between the rain forests and tropical grasslands.

tropopause The upper boundary of the troposphere.

troposphere The home of the biosphere; the lowest layer of the homosphere, containing approximately 90% of the total mass of the atmosphere; extends up to the tropopause, marked by a temperature of $-57°C$ ($-70°F$); occurring at an altitude of 18 km (11 mi) at the equator, 13 km (8 mi) in the middle latitudes, and near the poles at lower altitudes.

true north (or south) See *geographic North (or South) Pole*.

true shape A map property that shows the correct configuration of coastlines; a useful trait of conformality for navigational and aeronautical maps, although areal relationships are distorted.

typhoon A tropical cyclone with inward-spiraling winds in excess of 65 knots (74 mph) that occurs in the western Pacific; same as a hurricane except for location.

Universal Transverse Mercator (UTM) grid A special cylindrical projection in which the projection surface is tangent to the globe at a selected pair of opposing meridians or intersects at two selected small circles; scale is constant along those vertical lines. The UTM grid has become the standard for military applications.

vector A GIS data format where point, line, or polygon files are composed of coordinate pairs of x, y locations.

wilting point That point in the soil-moisture balance when only hygroscopic water and some bound capillary water remain. Plants wilt and eventually die after prolonged stress from a lack of available water.

wind-chill factor The enhanced rate at which body heat is lost to the air. As wind speeds increase, heat loss from the skin increases.

zenith The highest point reached by the Sun during its daily path.